let's talk science

# Adventures in the Scientific Method

level 4

**Author:** Carrie Lindquist

**Master Books Creative Team:**

**Editor:** Willow Meek

**Cover and Interior Design:**
Diana Bogardus
Terry White

**Copy Editors:**
Judy Lewis
Willow Meek

**Curriculum Review:**
Laura Welch
Diana Bogardus
Kristen Pratt

First printing: August 2022



Master Books, P.O. Box 726, Green Forest, AR 72638
Master Books® is a division of the New Leaf Publishing Group, Inc.

ISBN: 978-1-68344-289-9
ISBN: 978-1-61458-797-2 (digital)



Printed in the United States of America.

Please visit our website for other great titles: www.masterbooks.com



## About the Author

**Carrie Lindquist** is a homeschool graduate, wife to Wayne, and momma to three energetic boys. She is a passionate advocate for homeschooling and loves helping new-to-homeschooling moms realize that homeschooling through the early years isn't scary — it's really just an extension of all the fun things they are already doing with their children! When she isn't cleaning the endless little messes her boys create, you can find her encouraging moms to embrace the calling of everyday faithfulness.

# Table of Contents

# Course Description

Approximately 30 minutes per lesson, five times per week
Designed for fourth graders in a one-year course

In *Let's Talk Science Level 4: Adventures in the Scientific Method,* students will join characters Hannah and Ben as they explore the scientific method through the fields of chemistry and marine biology. Through conversational lessons, hands-on experimentation, observation, and documentation, students encounter God's design as they explore matter, elements, the carbon cycle, measurements, worldview, classification, the ocean, tide pools, coral reefs, algae, seagrass, and marine life.

As students discover God's wisdom in chemistry and marine biology, they'll also compile a personalized Science Notebook to document and share what they've learned each week. Along the way, they'll discover hidden treasures that teach them more about their relationship with God as they encounter God's wisdom, power, and majesty on display in creation.

# Course Objectives

Students completing this course will:

- Explore elements of chemistry and marine biology, including the scientific method, lab reports, measurements, matter, the periodic table of elements, mixtures, solutions, chemistry application, tides, tide pools, algae, seagrass, coral reefs, and marine life.
- Uncover how our worldview impacts the study of science and learn a simple question to discern observational science from historical science.
- Learn how to apply the scientific method and create their own lab reports.
- Discover that science teaches many lessons about God and our relationship with Him.
- Compile a unique Science Notebook as they document what they've learned and share it with others.

# A Note from the Author

Welcome to *Let's Talk Science Level 4: Adventures in the Scientific Method*! As your child begins this course, it's my prayer that they encounter the wisdom, power, majesty, and the grace of God in a new way this year as they explore His creation through chemistry and marine biology.

Each level of the *Let's Talk Science* series has been inspired by my sons and our learning adventures together. As they've continued to learn and grow, they've begun to realize that not everyone shares their worldview. This, of course, is especially true in science! This book was inspired by the questions they've asked and the topics we've explored together as they have learned how to discern worldview and apply biblical truth to what they are learning.

Science is fascinating and awe-inspiring because God's wisdom, understanding, majesty, and grace is so clearly on display in His creation. When we see His hand at work in the intricacies of atoms and molecules, the order of the elements, and even in His wise, merciful design for a fallen creation, it draws our hearts to worship Him.

*Let's Talk Science: Adventures in the Scientific Method* is more than an exploration of chemistry, marine biology, worldview, and the scientific method — it's an exploration of God's awe-inspiring creation, a discovery of the depth of His wisdom, a reminder that He is the source of truth, and an illustration of many spiritual lessons. Of course, sometimes science also reminds us that the world was broken through sin, and there we find God's mercy and a reminder that we all need Jesus.

This course was founded on a desire to teach my children to actively look for God's design as they study science — a skill they can take with them as they continue their education. *Let's Talk Science: Adventures in the Scientific Method* is designed to be relational and encourage curiosity — if your child is particularly interested in a topic or question, I invite you to spend some time exploring God's world together through books, videos, and resources. See if you can find additional aspects of God's design together. And don't forget, make this course your own — have fun!

It's my prayer that your student discovers God's amazing design, wisdom, and mercy in creation this year and that they will build their lives upon the firm foundation of His Word. May God richly bless your school year!

# Course Overview and Components

**Schedule:** The suggested weekly schedule is five days a week and has easy-to-manage lessons that guide the reading, worksheets, and activities. Teachers are encouraged to adjust the schedule as needed in order to best work within their unique educational program.

**Conversational Lessons:** The daily lessons in *Let's Talk Science: Adventures in the Scientific Method* are short and conversational. This allows confident readers to transition to student-led, independent instruction. If the student is younger or developing confidence in independent reading, the lessons may be read together or to the student.

**Science Notebook:** Students will create a personal Science Notebook to record what they've learned as they complete the course. The Science Notebook will be created in an artist sketchbook, which can be purchased in a craft store or online. It is recommended to purchase one with a sturdy cover, and it should contain a minimum of 36 blank pages.

Each week, the student will use their individual creativity to write and draw on one page of the Notebook. This hands-on expression allows the student to "own" what they've learned and share it in a personalized way. The student may use any medium (colored pencils, markers, paint, etc.) they prefer to complete the Notebook assignment.

Encourage your student to share their Science Notebook with friends and family as they tell what they've learned. Once the course is completed, the student's Science Notebook may also be saved as a keepsake.

**Materials List:** Provides an at-a-glance view of the supplies your student will need to complete course activities, as well as the week those supplies will be required.

**Vocabulary:** New vocabulary words will be introduced to the student in a **bold green font**. A phonetic pronunciation guide is also shown to help the student read the new word, such as in this example:

**environment** (said this way: ĕn-vī-rŭn-mĕnt)

For some words, the pronunciation may vary slightly depending upon the region in which you live. A vowel pronunciation key is included in the back of the book. Vocabulary words are also included in a glossary at the back of this book.

**Apply it:** This section features worksheets or activities to help the student apply what they've learned.

**Digging Deeper:** This section encourages the student to look up verses in their own Bible and memorize them. There are also occasional bonus questions, activities, or resources to help the student explore the topic deeper individually or as a family.

**Hidden Treasure:** Science illustrates and teaches us many things about our relationship with God. When we learn about our relationship with God through science, it's like finding a hidden treasure that we can share with others! This section is an opportunity for the student to share the Bible verse they learned during the week in their Science Notebook.

This component may be adjusted to fit the student's individual level. Older students may copy the verse on the back page of their Science Notebook by themselves, while younger students can help say the verse as the teacher writes it for them.

**Model Biomes:** After the student has finished exploring tide pools, coral reefs, and the Arctic Ocean, they will put together a model of that biome in a shoe or craft box. The lessons in weeks 23, 27, and 35 provide an instructional guide to create a simple, inexpensive biome. However, you may also customize and personalize the model biome as preferred. Many craft and hobby stores have model animals, trees, grass, and plants.

**Periodic Table of Elements:** The periodic table of elements is included in the back of this book for your student's reference. The student will need to refer to the periodic table in order to complete several Apply It assignments. It is recommended to laminate this page and store it in a safe place.

## Helpful Tips

### Tips for completing the Science Notebook:

- If you do not have a shoebox available, many craft and hobby stores sell shoebox-sized photo storage boxes. These are perfect for model biomes!
- Keep a variety of art supplies on hand, such as pencils, paint, colored pencils, markers, or crayons. Some students find watercolor pencils particularly enjoyable to use.
- Encourage creativity. Your student's artwork does not need to look exactly like any of the examples. The goal of the Science Notebook is for the student to express what they've learned in a way that is unique and personalized. This develops the student's ability to "own" what they've learned about.
- If your student is concerned about making mistakes, it can be helpful for them to begin drawing with a pencil. This allows them to erase perceived mistakes. Once they're done drawing, they can add colors and/or additional details.
- Encourage enjoyment rather than perfection. Three different students' Science Notebooks are included with each lesson to show different ways the student can draw the prompt. Point out that each Notebook is unique and shows the creativity God has given that student.
- If your student is reluctant to draw, it can be helpful to sit down with them and draw together. Point out a basic shape you can start with, such as a circle, oval, or rectangle, then add additional details.

**Additional Tips:**

- Lesson 5 covers a brief overview of the imperial system and the metric system. This lesson is intended to provide a brief and basic overview to prepare the student for Lesson 6. *Math Lessons for a Living Education Level 4* will cover the metric system at a deeper level beginning in Lesson 15.
- In Lesson 17, students learn to ask, "Can this information be observed, tested, and repeated?" to help them determine if they are learning about historical or observational science. As your student reads science books or watches television, encourage them to ask and answer this question and talk about the information they are learning.
- Lesson 19 is an introduction to taxonomy. This lesson is intended to provide a brief and basic overview of how living things are classified.
- *Let's Talk Science: Adventures in the Scientific Method* does touch on difficult topics such as animal death in creation. It is encouraged to discuss these topics deeper with your student, depending on their individual level.

**Recommended Resources from Master Books®:**

***The Answers Book for Kids Volume 3:*** Prompts for the student to read more about where the Bible came from and about worldview in this resource are included in Lessons 16 and 17. This book is not required to complete the course.

***Dragons of the Deep:*** Your student may enjoy exploring the marine creatures in this book after Lesson 33. This book is not required to complete the course.

***The New Ocean Book:*** The student may use this resource to look up topics of interest in order to learn about the ocean, marine life, and ocean exploration at a deeper level. This book is not required to complete the course.

# Master Materials List

**Basic Supplies:**

These supplies will be used often throughout the course.

- ☐ Artist sketchbook (36 pages minimum)
- ☐ Colored pencils, markers, acrylic paint set with paintbrush, crayons
- ☐ Permanent marker
- ☐ Silver or white glitter glue
- ☐ Glue stick
- ☐ Stapler
- ☐ Scissors
- ☐ Ruler
- ☐ Construction paper
- ☐ 4 shoeboxes or craft boxes with lids
- ☐ Plastic tablecloths
- ☐ Measuring cups, tablespoon, teaspoon
- ☐ Toothpicks

**Week 2**

- ☐ Vanilla extract
- ☐ Plate

**Week 3**

- ☐ Baking soda
- ☐ Vinegar
- ☐ Lemon juice
- ☐ 2 bowls

**Week 4**

- ☐ 2 balloons
- ☐ 2 soda bottles (same size)
- ☐ Vinegar
- ☐ Lemon juice
- ☐ Baking soda
- ☐ Funnel
- ☐ Soft tailor's tape measure

**Week 5**

- ☐ Measuring tape
- ☐ Bathroom scale
- ☐ Gallon container

**Week 6**

- ☐ Kitchen scale that measures in grams or kilograms
- ☐ 2 balloons
- ☐ Yarn

**Week 7**

- ☐ An apple or orange
- ☐ Knife (adult supervision)
- ☐ Kitchen scale
- ☐ Cutting board

**Week 8**

- ☐ 2 bar magnets
- ☐ Plain M&M'S®

**Week 11**

- ☐ Marshmallows
- ☐ Dirt
- ☐ 2 bowls
- ☐ Hydrogen peroxide

**Week 12**

- ☐ Marshmallows

**Week 13**

- ☐ 2 clear glasses or coffee mugs
- ☐ 2 tea bags
- ☐ Sugar
- ☐ Olive or coconut oil
- ☐ Optional: milk
- ☐ Sticky tabs
- ☐ Salt
- ☐ Stopwatch

**Week 14**

- ☐ pH test kit or meter for soil
- ☐ Local soil

**Week 15**

- ☐ Sunglasses

**Week 20**

- ☐ Kitchen scale that measures in grams
- ☐ Pitcher
- ☐ Salt
- ☐ Jar or container
- ☐ Plate
- ☐ Ocean water from Day 2

### Week 21

- ☐ Glass or plastic bottle with screw-on lid
- ☐ Blue food coloring
- ☐ Coconut or vegetable oil
- ☐ Funnel

### Week 22

- ☐ Air dry modeling clay
- ☐ Paper plate

### Week 23

- ☐ Air dry modeling clay
- ☐ Paper plates
- ☐ Hot glue gun (adult only)
- ☐ 3 red craft pipe cleaners
- ☐ 2 googly craft eyes
- ☐ Rocks or sand
- ☐ Blue construction paper
- ☐ Craft model sea star, sea urchins, clams, and crab from previous lessons

### Week 25

- ☐ 6 Styrofoam™ cups

### Week 26

- ☐ Air dry modeling clay
- ☐ Yellow or brown acrylic paint
- ☐ Paper plate

### Week 27

- ☐ Toilet paper roll
- ☐ Colored tissue paper
- ☐ Sand
- ☐ Craft model brain coral and anemones from previous lessons
- ☐ Small model marine life such as turtle, manta ray, fish, etc.
- ☐ Optional: hot glue gun (adult only)

### Week 28

- ☐ Stopwatch

### Week 29

- ☐ Blindfold
- ☐ Clothespin
- ☐ Toothbrush
- ☐ Pepper
- ☐ Kitchen scale

### Week 32

- ☐ Modeling clay

### Week 33

- ☐ Onion
- ☐ Knife (adult supervision)
- ☐ Cutting board

### Week 34

- ☐ Ice cubes
- ☐ Drinking glass
- ☐ Light and dark blue acrylic paint

### Week 35

- ☐ Cotton balls
- ☐ Utility knife (adult supervision)
- ☐ Clear or silver glitter glue
- ☐ ½–1 inch thick Styrofoam™ panel or block
- ☐ Small model arctic marine mammals such as polar bear, seal, beluga, narwhale, or bowhead whale

## Schedule

| Date | Day | Assignment | Due Date | ✓ | Grade |
|---|---|---|---|---|---|
| Week 1 | Day 1 | The Adventure Begins • Complete reading and activity on pages 19–20 | | | |
| | Day 2 | Complete reading and activity on pages 21–22 | | | |
| | Day 3 | Complete reading and activity on pages 23–24 | | | |
| | Day 4 | Complete reading and activity on page 25 | | | |
| | Day 5 | Complete reading and activity in your Notebook on pages 26–28 | | | |
| Week 2 | Day 6 | Science Through History • Complete reading and activity on pages 29–30 | | | |
| | Day 7 | Complete reading and activity on pages 31–32 | | | |
| | Day 8 | Complete reading and activity on pages 33–34 | | | |
| | Day 9 | Complete reading and activity on page 35 | | | |
| | Day 10 | Complete reading and activity in your Notebook on page 36 | | | |
| Week3 | Day 11 | The Scientific Method • Complete reading and activity on pages 37–38 | | | |
| | Day 12 | Complete reading and activity on pages 39–40 | | | |
| | Day 13 | Complete reading and activity on pages 41–42 | | | |
| | Day 14 | Complete reading and activity on pages 43–44 | | | |
| | Day 15 | Complete reading and activity in your Notebook on pages 45–46 | | | |
| Week 4 | Day 16 | Lab Reports • Complete reading and activity on pages 47–48 | | | |
| | Day 17 | Complete reading and activity on pages 49–52 | | | |
| | Day 18 | Complete reading and activity on pages 53–54 | | | |
| | Day 19 | Complete reading and activity on pages 55–56 | | | |
| | Day 20 | Complete reading and activity in your Notebook on pages 57–58 | | | |
| Week 5 | Day 21 | Systems of Measurements • Complete reading and activity on pages 59–60 | | | |
| | Day 22 | Complete reading and activity on pages 61–62 | | | |
| | Day 23 | Complete reading and activity on pages 63–64 | | | |
| | Day 24 | Complete reading and activity on pages 65–66 | | | |
| | Day 25 | Complete reading and activity in your Notebook on pages 67–68 | | | |

# Schedule

| Date | Day | Assignment | Due Date | ✓ | Grade |
|---|---|---|---|---|---|
| Week 6 | Day 1 | Mass & Matter • Complete reading and activity on pages 69–70 | | | |
| | Day 2 | Complete reading and activity on pages 71–72 | | | |
| | Day 3 | Complete reading and activity on pages 73–76 | | | |
| | Day 4 | Complete reading and activity on pages 77–78 | | | |
| | Day 5 | Complete reading and activity in your Notebook on pages 79–80 | | | |
| Week 7 | Day 1 | Describing Matter • Complete reading and activity on pages 81–82 | | | |
| | Day 2 | Complete reading and activity on pages 83–84 | | | |
| | Day 3 | Complete reading and activity on pages 85–86 | | | |
| | Day 4 | Complete reading and activity on pages 87–88 | | | |
| | Day 5 | Complete reading and activity in your Notebook on pages 89–90 | | | |
| Week 8 | Day 1 | Building Blocks of Matter • Complete reading and activity on pages 91–92 | | | |
| | Day 2 | Complete reading and activity on pages 93–94 | | | |
| | Day 3 | Complete reading and activity on page 95 | | | |
| | Day 4 | Complete reading and activity on page 96 | | | |
| | Day 5 | Complete reading and activity in your Notebook on pages 97–98 | | | |
| Week 9 | Day 1 | Elements • Complete reading and activity on pages 99–100 | | | |
| | Day 2 | Complete reading and activity on pages 101–102 | | | |
| | Day 3 | Complete reading and activity on pages 103–104 | | | |
| | Day 4 | Complete reading and activity on pages 105–106 | | | |
| | Day 5 | Complete reading and activity in your Notebook on pages 107–108 | | | |
| Week 10 | Day 1 | Periodic Table of Elements • Complete reading and activity on pages 109–110 | | | |
| | Day 2 | Complete reading and activity on pages 111–112 | | | |
| | Day 3 | Complete reading and activity on pages 113–114 | | | |
| | Day 4 | Complete reading and activity on page 115 | | | |
| | Day 5 | Complete reading and activity in your Notebook on page 116 | | | |

## Schedule

| Date | Day | Assignment | Due Date | ✓ | Grade |
|---|---|---|---|---|---|
| Week 11 | Day 1 | Molecules • Complete reading and activity on pages 117–118 | | | |
| | Day 2 | Complete reading and activity on pages 119–122 | | | |
| | Day 3 | Complete reading and activity on pages 123–124 | | | |
| | Day 4 | Complete reading and activity on page 125 | | | |
| | Day 5 | Complete reading and activity in your Notebook on page 126 | | | |
| Week 12 | Day 1 | The Carbon Cycle • Complete reading and activity on pages 127–128 | | | |
| | Day 2 | Complete reading and activity on pages 129–130 | | | |
| | Day 3 | Complete reading and activity on pages 131–132 | | | |
| | Day 4 | Complete reading and activity on pages 133–134 | | | |
| | Day 5 | Complete reading and activity in your Notebook on pages 135–136 | | | |
| Week 13 | Day 1 | Mixtures & Solutions • Complete reading and activity on pages 137–138 | | | |
| | Day 2 | Complete reading and activity on pages 139–140 | | | |
| | Day 3 | Complete reading and activity on pages 141–144 | | | |
| | Day 4 | Complete reading and activity on page 145 | | | |
| | Day 5 | Complete reading and activity in your Notebook on page 146 | | | |
| Week 14 | Day 1 | Chemistry Around Us • Complete reading and activity on pages 147–148 | | | |
| | Day 2 | Complete reading and activity on pages 149–150 | | | |
| | Day 3 | Complete reading and activity on pages 151–154 | | | |
| | Day 4 | Complete reading and activity on page 155 | | | |
| | Day 5 | Complete reading and activity in your Notebook on page 156 | | | |
| Week 15 | Day 1 | Worldview 1 • Complete reading and activity on pages 157–158 | | | |
| | Day 2 | Complete reading and activity on pages 159–160 | | | |
| | Day 3 | Complete reading and activity on pages 161–162 | | | |
| | Day 4 | Complete reading and activity on page 163 | | | |
| | Day 5 | Complete reading and activity in your Notebook on page 164 | | | |

## Schedule

| Date | Day | Assignment | Due Date | ✓ | Grade |
|---|---|---|---|---|---|
| Week 16 | Day 1 | Worldview 2 • Complete reading and activity on pages 165–166 | | | |
| | Day 2 | Complete reading and activity on pages 167–168 | | | |
| | Day 3 | Complete reading and activity on pages 169–170 | | | |
| | Day 4 | Complete reading and activity on page 171 | | | |
| | Day 5 | Complete reading and activity in your Notebook on page 172 | | | |
| Week 17 | Day 1 | Observational & Historical Science • Complete reading and activity on pages 173–174 | | | |
| | Day 2 | Complete reading and activity on pages 175–176 | | | |
| | Day 3 | Complete reading and activity on pages 177–178 | | | |
| | Day 4 | Complete reading and activity on page 179 | | | |
| | Day 5 | Complete reading and activity in your Notebook on page 180 | | | |
| Week 18 | Day 1 | Living Things • Complete reading and activity on pages 181–182 | | | |
| | Day 2 | Complete reading and activity on pages 183–184 | | | |
| | Day 3 | Complete reading and activity on pages 185–186 | | | |
| | Day 4 | Complete reading and activity on pages 187–188 | | | |
| | Day 5 | Complete reading and activity in your Notebook on pages 189–190 | | | |
| Week 19 | Day 1 | Classification • Complete reading and activity on pages 191–193 | | | |
| | Day 2 | Complete reading and activity on pages 195–196 | | | |
| | Day 3 | Complete reading and activity on pages 197–198 | | | |
| | Day 4 | Complete reading and activity on page 199 | | | |
| | Day 5 | Complete reading and activity in your Notebook on page 200 | | | |
| Week 20 | Day 1 | Marine Biology • Complete reading and activity on pages 201–202 | | | |
| | Day 2 | Complete reading and activity on pages 203–204 | | | |
| | Day 3 | Complete reading and activity on pages 205–208 | | | |
| | Day 4 | Complete reading and activity on page 209 | | | |
| | Day 5 | Complete reading and activity in your Notebook on page 210 | | | |

## Schedule

| Date | Day | Assignment | Due Date | ✓ | Grade |
|---|---|---|---|---|---|
| Week 21 | Day 1 | Ocean Movement • Complete reading and activity on pages 211–212 | | | |
| | Day 2 | Complete reading and activity on pages 213–214 | | | |
| | Day 3 | Complete reading and activity on pages 215–216 | | | |
| | Day 4 | Complete reading and activity on page 217 | | | |
| | Day 5 | Complete reading and activity in your Notebook on page 218 | | | |
| Week 22 | Day 1 | Tide Pools 1 • Complete reading and activity on pages 219–220 | | | |
| | Day 2 | Complete reading and activity on pages 221–222 | | | |
| | Day 3 | Complete reading and activity on pages 223–224 | | | |
| | Day 4 | Complete reading and activity on page 225 | | | |
| | Day 5 | Complete reading and activity in your Notebook on page 226 | | | |
| Week 23 | Day 1 | Tide Pools 2 • Complete reading and activity on pages 227–228 | | | |
| | Day 2 | Complete reading and activity on pages 229–230 | | | |
| | Day 3 | Complete reading and activity on pages 231–232 | | | |
| | Day 4 | Complete reading and activity on page 233 | | | |
| | Day 5 | Complete reading and activity in your Notebook on page 234 | | | |
| Week 24 | Day 1 | Algae & Seagrass • Complete reading and activity on pages 235–236 | | | |
| | Day 2 | Complete reading and activity on pages 237–238 | | | |
| | Day 3 | Complete reading and activity on pages 239–240 | | | |
| | Day 4 | Complete reading and activity on page 241 | | | |
| | Day 5 | Complete reading and activity in your Notebook on page 242 | | | |

## Schedule

| Date | Day | Assignment | Due Date | ✓ | Grade |
|---|---|---|---|---|---|
| Week 25 | Day 1 | Marine Food Chain • Complete reading and activity on pages 243–244 | | | |
| | Day 2 | Complete reading and activity on pages 245–246 | | | |
| | Day 3 | Complete reading and activity on pages 247–249 | | | |
| | Day 4 | Complete reading and activity on page 251 | | | |
| | Day 5 | Complete reading and activity in your Notebook on page 252 | | | |
| Week 26 | Day 1 | The Coral Reef • Complete reading and activity on pages 253–254 | | | |
| | Day 2 | Complete reading and activity on pages 255–256 | | | |
| | Day 3 | Complete reading and activity on pages 257–258 | | | |
| | Day 4 | Complete reading and activity on page 259 | | | |
| | Day 5 | Complete reading and activity in your Notebook on page 260 | | | |
| Week 27 | Day 1 | The Great Barrier Reef • Complete reading and activity on pages 261–262 | | | |
| | Day 2 | Complete reading and activity on pages 263–264 | | | |
| | Day 3 | Complete reading and activity on pages 265–266 | | | |
| | Day 4 | Complete reading and activity on page 267 | | | |
| | Day 5 | Complete reading and activity in your Notebook on page 268 | | | |
| Week 28 | Day 1 | Whales 1 • Complete reading and activity on pages 269–270 | | | |
| | Day 2 | Complete reading and activity on pages 271–272 | | | |
| | Day 3 | Complete reading and activity on pages 273–274 | | | |
| | Day 4 | Complete reading and activity on pages 275–276 | | | |
| | Day 5 | Complete reading and activity in your Notebook on pages 277–278 | | | |
| Week 29 | Day 1 | Whales 2 • Complete reading and activity on pages 279–280 | | | |
| | Day 2 | Complete reading and activity on pages 281–284 | | | |
| | Day 3 | Complete reading and activity on pages 285–286 | | | |
| | Day 4 | Complete reading and activity on pages 287–288 | | | |
| | Day 5 | Complete reading and activity in your Notebook on pages 289–290 | | | |

## Schedule

| Date | Day | Assignment | Due Date | ✓ | Grade |
|---|---|---|---|---|---|
| Week 30 | Day 1 | Conservation • Complete reading and activity on pages 291–292 | | | |
| | Day 2 | Complete reading and activity on pages 293–294 | | | |
| | Day 3 | Complete reading and activity on pages 295–296 | | | |
| | Day 4 | Complete reading and activity on page 297 | | | |
| | Day 5 | Complete reading and activity in your Notebook on page 298 | | | |
| Week 31 | Day 1 | Scientific Method in Marine Biology • Complete reading and activity on pages 299–300 | | | |
| | Day 2 | Complete reading and activity on pages 301–302 | | | |
| | Day 3 | Complete reading and activity on pages 303–304 | | | |
| | Day 4 | Complete reading and activity on pages 305–306 | | | |
| | Day 5 | Complete reading and activity in your Notebook on pages 307–308 | | | |
| Week 32 | Day 1 | Sharks 1 • Complete reading and activity on pages 309–310 | | | |
| | Day 2 | Complete reading and activity on pages 311–312 | | | |
| | Day 3 | Complete reading and activity on pages 313–314 | | | |
| | Day 4 | Complete reading and activity on page 315 | | | |
| | Day 5 | Complete reading and activity in your Notebook on page 316 | | | |
| Week 33 | Day 1 | Sharks 2 • Complete reading and activity on pages 317–320 | | | |
| | Day 2 | Complete reading and activity on pages 321–322 | | | |
| | Day 3 | Complete reading and activity on pages 323–324 | | | |
| | Day 4 | Complete reading and activity on pages 325–326 | | | |
| | Day 5 | Complete reading and activity in your Notebook on pages 327–328 | | | |
| Week 34 | Day 1 | Arctic Ocean 1 • Complete reading and activity on pages 329–330 | | | |
| | Day 2 | Complete reading and activity on pages 331–334 | | | |
| | Day 3 | Complete reading and activity on pages 335–336 | | | |
| | Day 4 | Complete reading and activity on page 337 | | | |
| | Day 5 | Complete reading and activity in your Notebook on page 338 | | | |

## Schedule

| Date | Day | Assignment | Due Date | ✓ | Grade |
|---|---|---|---|---|---|
| Week 35 | Day 1 | Arctic Ocean 2 • Complete reading and activity on pages 339–340 | | | |
| | Day 2 | Complete reading and activity on pages 341–342 | | | |
| | Day 3 | Complete reading and activity on pages 343–344 | | | |
| | Day 4 | Complete reading and activity on page 345 | | | |
| | Day 5 | Complete reading and activity in your Notebook on page 346 | | | |
| Week 36 | Day 1 | Review • Complete reading and activity on pages 347–348 | | | |
| | Day 2 | Complete reading and activity on pages 349–350 | | | |
| | Day 3 | Complete reading and activity on pages 351–352 | | | |
| | Day 4 | Complete reading and activity on pages 353–354 | | | |
| | Day 5 | Complete reading and activity in your Notebook on pages 355–356 | | | |

week 1

# The Adventure Begins

Day 1

Hello there! We're so glad you're here; we've been waiting for you! My name is Ben, and this is my older sister, Hannah. Hannah and I love to explore the world God created through science.

That's right! Science is a tool God has given us. When we study God's creation through science, it helps us to ask questions, test our ideas, and share what we've learned with others. As we learn, we also discover more about God and our relationship with Him.

Ben and I have had a lot of science adventures together — in fact, we're getting ready to start a new one right now! We are excited that you're going to join us on our new adventure. Are you ready to get started?

Whoa, whoa, whoa, Hannah — we haven't even told our friend what our science adventure is going to be about this year!

Oh, right — I guess I got a little excited. Well, this year, Ben and I want to dive deeper into what science is. We have some big questions like: How does science help us to ask questions? How does science guide us as we test our ideas? Can we trust science? How can we use science to help us learn more about God's creation?

But that's not all! We also want to learn about some real scientists. We want to know what they believed, what questions they asked, and all about what they discovered. Our mom told us that we're going to answer all of our questions as we explore God's creation through biology and chemistry this year.

Name: ______

I can hardly wait! Our mom also told us that we're going to do plenty of activities as we answer our questions. We have a lot to learn about, so where should we start, Hannah?

Hmm, we will start our adventure this week by laying a good foundation. There are a few words and definitions we'll need to learn before we get started on our science adventure. Remember, a **definition** (said this way: dĕf-ŭh-nĭsh-ŭhn) is what a word means.

I hear Mom calling us now, which means we're out of time for today. Let's plan on talking more about what science is tomorrow — we'll need to go deeper than we ever have before!

apply it

1. What is something you have enjoyed learning about in science in the past?

2. What is something you would like to learn about in science?

3. What does the word "definition" mean?

Welcome back, friend! We're going to begin our adventure by taking a deeper look at the word "science."

But Ben, science isn't really a new word for us. We've been learning about science for years! Why do we need to start here?

You're right, Hannah, but now it's time to take a deeper look at what science is.

Okay, well, to begin, we know that science is a tool God has given us. When we study God's creation through science, it helps us to ask questions, test our ideas, and share what we've learned with others. As we explore, we also learn more about God and our relationship with Him.

That is our foundation for what science is. Now let's build on it! Hmm, I have a question for you: what do we have inside of our bodies that helps us to stand up straight and tall?

I know! Our bones, or skeleton!

Yes! Our skeleton gives our body the structure it needs to stand and move around. Without our skeleton, we'd be all floppy like a jellyfish.

**Structure** (said this way: strŭk-cher) provides a way to organize or support something. Science is a tool that gives us the structure we need to explore the world in an organized way — kind of like how our skeleton supports our body.

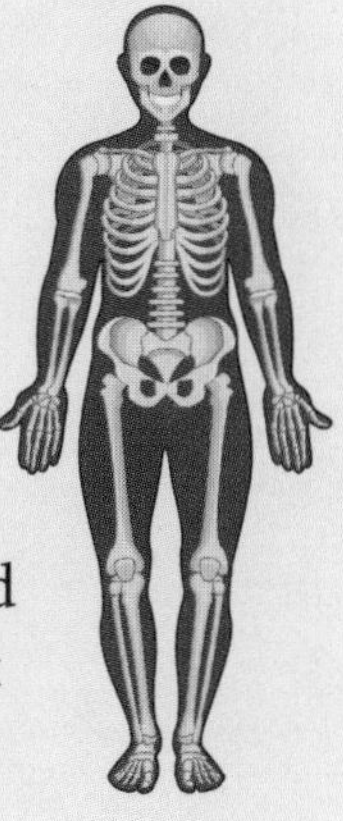

God gave us the ability to learn and explore. We can grow our knowledge and understanding about the world God created by studying science. The structure that science gives us helps us to know where to begin, how to test our ideas, and how to share our discoveries with others. Science helps us to pursue knowledge and understanding about God's creation. **Pursue** (said this way: per-soo) is a word that means to chase after, seek, or search for something.

Okay! So what you're saying is that **science** is the pursuit of knowledge and understanding about God's creation through an organized process.

Name:

Exactly! Let's take that definition now and add what we already knew about science. Here is our definition of science for this year:

> "Science is the pursuit of knowledge and understanding about God's creation through an organized process. Science helps us to ask questions, test our ideas, and share what we've learned with others. Through science, we also learn more about God and our relationship with Him."

We're going to be talking much more about the organized process science provides in the weeks to come. In the meantime, let's make sure we understand what the words "knowledge," "understanding," and "organize" mean!

1. Circle these key words in our definition of science below: knowledge, understanding, creation, ask, test, share, God

   Science is the pursuit of knowledge and understanding about God's creation through an organized process. Science helps us to ask questions, test our ideas, and share what we've learned with others. Through science, we also learn more about God and our relationship with Him.

2. Ask your teacher to help you look up each word in a dictionary. Copy the definition of each word.

**Knowledge:**

**Understanding:**

**Organize:**

In our last lesson, we explored what science is. Let's review what we know so far:

> "Science is the pursuit of knowledge and understanding about God's creation through an organized process. Science helps us to ask questions, test our ideas, and share what we've learned with others. Through science, we also learn more about God and our relationship with Him."

One important word in our definition is the word "organized." Science is all about organization. It helps us to organize how we study the world around us, and it even helps us to organize all the things we learn about. Even science itself is organized!

Oh, that's right! There are many things we can explore in God's creation like plants, animals, the ocean, machines, the universe, chemicals, and so much more! Each of these different things is organized into a field, or branch, of science.

When a scientist studies the sun, moon, stars, or the far reaches of the universe, that scientist is studying the field of astronomy. We call the scientist an astronomer. When a scientist explores plants and animals, they are exploring the field of biology. We would call that scientist a biologist.

This organization helps a scientist to be able to focus in on what they are studying. For example, rather than study the whole entire world, a biologist can choose to focus on studying plants. The biologist can then work toward learning everything he or she can about several types of plants or even just one specific plant.

Science helps us to stay organized and keep our focus on what we're learning about. This year, we're going to focus on exploring God's creation through biology and chemistry.

Ooh, I'm excited now! The field of biology explores the different types of life God created, from plants to people, and even animals. In chemistry, we discover the building blocks that make everything we see, from living things to non-living things.

As we explore biology and chemistry together, we'll also be learning about the scientific method. The scientific method is how we organize our study of science. We'll learn more about the scientific method and who developed it in the coming weeks!

Name: ____________________

apply it

1. Do you remember our definition of science? Fill in the missing words. You can look back in the lesson if you need to review!

"Science is the pursuit of ____________________ and ____________________ about God's ____________________ through an organized process. Science helps us to ______________ questions, ______________ our ideas, and ______________ what we've learned with others. Through science, we also learn more about ______________ and our relationship with Him."

2. What fields of science have you heard about?

3. Have you learned about a scientist in the past? If so, write their name and the field of science that they studied below.

As we've been talking about what science is this week, I've been thinking about one other reason we explore God's creation through science.

Oh? What is that, Ben?

Well, as we pursue knowledge and understanding of how God's creation works, we also catch glimpses of who God is. When we see how organized His creation is, it shows us that God as our Creator is also organized.

When we see one of His amazing designs in a plant, animal, or even a chemical, it reminds us that God is wise and powerful. When we learn more about God through science, we want to share that with the people around us!

It reminds me of a verse we read in Sunday school this week, Psalm 9:1–2:

*I will give thanks to you, LORD, with all my heart; I will tell of all your wonderful deeds. I will be glad and rejoice in you; I will sing the praises of your name, O Most High.*

I agree, Ben! When we learn more about science, it reminds me to worship and praise God. Seeing His wisdom and power on display in His creation makes me so thankful that He also cares for us. The more I learn, the more I want to tell others about His wonderful deeds! Hey, I have an idea, friend! Let's share Psalm 9:1–2 with a friend or our family and talk about God's wonderful deeds together.

**Note**

Day Four provides a spiritual connection for the student. There is not a worksheet on these days. Instead, the student will look up the verses and memorize from their Bible using the Digging Deeper section.

Look up Psalm 9:1–2 in your Bible. You can ask your teacher for help if you need to. If you'd like, you can highlight these verses in your Bible. Memorize Psalm 9:1–2 with your teacher or with a sibling.

Whew! We've made it to the end of our first week together — we've talked about a lot!

Let's see, we talked about what science is, and we created our definition of science. We also discussed how science is organized and how learning about science leads our hearts to praise God and tell others about Him.

And this is just the beginning! We have a lot more to learn, so we're going to need a way to document and share what we've learned with others.

Scientists throughout history have written about, drawn, or taken pictures of the things they've observed, tested, and learned so that they can share them with others. As we explore God's creation through biology and chemistry this year, we're going to create our own Science Notebook!

We'll record the things we learn about God's creation and the experiments we do in our Notebook. Each week you can share what you learned with someone else, just like a real scientist!

And my favorite part is that at the end of the year, we'll be able to look through our Notebook and remember all the things we discovered along the way! What should we add to the first page of our Science Notebook today, Hannah?

Let's write the first part of our science definition and draw a picture of the earth, one of God's creations! I have a picture of the earth right here that we can use to give us an idea for how to create our drawings.

Our younger brother Sam is too little to write the sentence with us, but he likes to add a new drawing to his Science Notebook each week just like we do. We'll show you how each of our Notebooks turns out — each of ours is unique. Have fun creating your drawing!

In your Science Notebook, write:
**Science is the pursuit of knowledge and understanding about God's creation through an organized process.**

Then draw a picture of the earth.

**Hidden Treasure** Psalm 9:1–2 reminded us this week that as we learn, we can praise God and tell others about His wonderful deeds. Copy Psalm 9:1–2 on the back of your Notebook page.

*I will give thanks to you, Lord, with all my heart; I will tell of all your wonderful deeds. I will be glad and rejoice in you; I will sing the praises of your name, O Most High* (Psalm 9:1–2).

# Science Through History

Day 1

Welcome back, friend! It's a new day, and we're ready for another science adventure together. Let's dive right in by talking about a little bit of history today. We can start at the very beginning — we know from Genesis chapter 1 in the Bible that God created the heavens and the earth. We also know that God created the first man and woman, Adam and Eve. In Genesis 1:27 we read,

*So God created mankind in his own image, in the image of God he created them; male and female he created them.*

Ah, I remember! We talked about what it means to be made in the image of God in *Adventures on Planet Earth*. Let's quickly review what we learned. To be made in the image of God doesn't mean that we look just like God. Instead, it means that we reflect the character and the attributes of God.

Remember, character is a word that means the features or traits of something or someone. An attribute is a quality of something or someone. We learn about God's character and attributes in the Bible. For example, we read in 1 John 4:8 that God is love. Deuteronomy 32:4 tells us that God is just. When we study creation, we also see some of the attributes of God such as His creativity and organization in what He created.

God created humans in His image to reflect His character and attributes. When we love each other, we're showing others God's loving character. We're reflecting God's love.

But remember, soon after creation, something changed. Adam and Eve chose to sin — to disobey God's direction. Sin always breaks and destroys. Before sin, people would have reflected the image of God perfectly. But once sin entered, it stained the perfect image of God in us. People still bear the image of God — but it is imperfect and broken now. When we are unloving, unjust, unmerciful, unkind, and ungracious, we reflect the image of sin rather than the image of God.

Thanks, Hannah! We're out of time for today, but tomorrow we'll continue to explore the image of God and how that affects the way we study science. See you then!

**Weekly materials list**

- Vanilla extract ☐
- Plate ☐

Name:

apply it

1. Read Genesis 3 with your teacher. What do you think that day was like?

2. How do you imagine Adam and Eve felt?

3. Draw a picture of what you think it looked like as they left the beautiful Garden of Eden.

I'm excited to continue our discussion today! Let's see . . . last time, we talked about the image of God, as well as His attributes and character. We find many of God's attributes in the Bible. As we study His creation, we also observe His attributes and character.

**materials needed**

- Vanilla extract
- Plate

Just think for a minute about all of the different plants and creatures we see around us — these display God's creativity! When we study the phases of the moon and the order of the planets, we are also seeing God's organization, power, and majesty on display.

God created people in His image, and He gave them the ability to be creative in many different ways. Some people make beautiful art, others create lovely music, some write exciting stories — there are so many different ways we can be creative! God also gave people the ability to think, organize, and solve problems. He gave us curiosity to learn more and a desire to understand the world around us.

Since creation, men and women have been driven to understand more about God's creation. We have wanted to understand how things work — and why they work the way they do. Many men and women have studied science all throughout history.

But there was a problem — there wasn't a consistent way for people to study the world around them. Some people studied creation through logic. Logic is what we use when we think, or reason, through a problem or question. Logic is a very helpful and important tool for studying science — but sometimes what we observe in creation doesn't make logical sense to us at first.

Ooh, that's right! I was reading about this just last night. For many years, people told stories about burning rocks that fell from the sky. Fiery rocks falling from the sky didn't make any sense to the scientists at that time — they thought these stories were just plain crazy! Later on, however, meteorites that crashed to earth from outer space were discovered, and finally the stories of burning rocks that fell from the sky made sense.

Other scientists relied on their five senses to help them answer questions. But there was a problem with that too — our five senses can sometimes be wrong or even deceive us.

Ah, like watching Uncle Gus' "magic" tricks! It may look like he can make something disappear, but it is really a trick on our eyes. Mom said she had an activity to help us test how trustworthy our senses are. Let's go have some fun!

Name: ______________________

## Activity directions:

1. Wash your hands then place a drop of vanilla extract on the plate.
2. Use your sense of sight, touch, and smell to decide what you think the vanilla will taste like. Write your answers for questions 1–4 on the worksheet below.
3. Dip your finger into the vanilla extract and then taste it — did the taste match what your other senses told you it would be like? Write your answers for questions 5–6 on the worksheet.

1. What does the vanilla extract look like?

______________________

______________________

2. What does the vanilla extract feel like?

______________________

______________________

3. What does the vanilla extract smell like?

______________________

______________________

4. Based on what you can see, feel, and smell, how do you think the vanilla extract will taste?

______________________

______________________

5. What did the vanilla extract taste like?

______________________

______________________

6. Is that what you expected?

______________________

______________________

What did you think of that vanilla extract yesterday, Ben?

It's good in cookies and cakes when Mom uses it to bake, but I don't want to taste it by itself ever again!

I agree! Our activity showed how our senses can be unreliable sometimes. Our sense of smell told us that the vanilla extract should taste wonderful — but our sense of taste showed us it was actually bitter and gross.

In the same way, when scientists through history relied the most on what they could see, hear, smell, taste, or see, their conclusions were often incorrect.

A **conclusion** (said this way: kŭhn-kloo-zhŭn) is a result or decision about something. For example, our conclusion about the vanilla extract was that though it smelled nice, it was really bitter.

I have another example from history of an incorrect conclusion. If you observe the sky, you may notice the sun travels across the sky as it rises and sets. You may also notice that the stars do not remain in the same place. Finally, you may realize that you are standing still — you cannot feel any movement under your feet. What do you think you might conclude about the sun and stars from what you can see and feel?

In the past, scientists made these observations and concluded that everything in space revolved, or orbited, around the earth. They concluded that the earth was at the center of our solar system.

Their observation made logical sense based on what they could see and feel — but it wasn't really how things worked. This theory was called **geocentrism** (said this way: jē-ō-sĕn-trĭsm), and it was the common scientific belief for over 1,000 years!

Over time, scientists began noticing problems with geocentrism — there were questions it couldn't answer, and some things just didn't make sense. Nicolaus Copernicus built on the work of other scientists to develop the theory that the sun was really at the center of the solar system and the earth and other planets orbited around it. His theory is called **heliocentrism** (said this way: hē-lē-ō-sĕn-trĭsm).

Name: ______________________________

*Geo* refers to the earth, and *helio* refers to the sun. Geocentrism means "earth centered," and heliocentrism means "sun centered."

Other scientists like Galileo Galilei and Johann Kepler worked to gather evidence to support heliocentrism. Though it took a long time for people to accept this new theory, heliocentrism is how we understand the orbits of the earth and planets today.

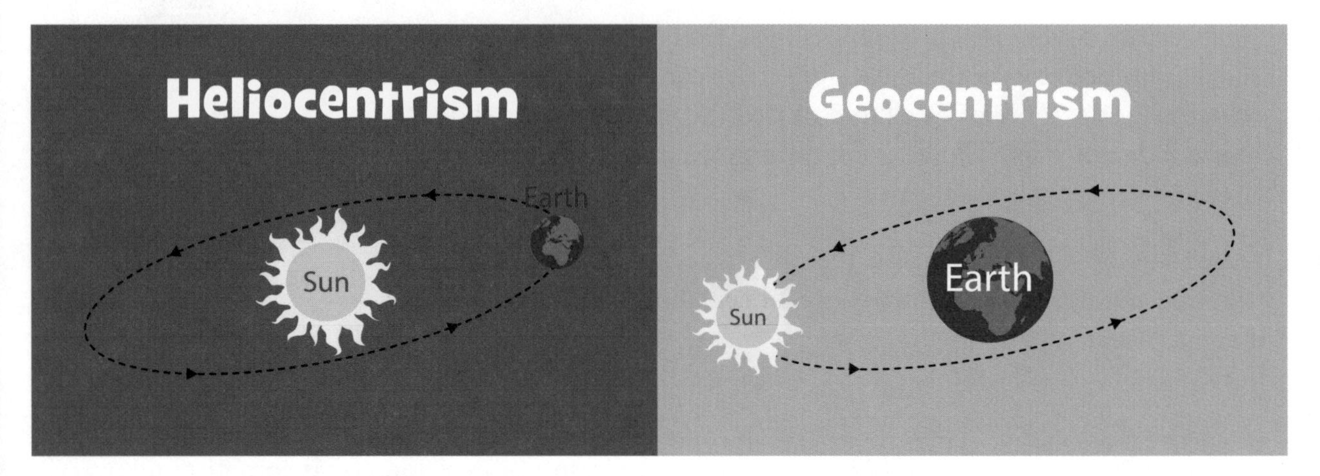

1. Match each word to the correct definition.

| | |
|---|---|
| **Heliocentrism** | The theory that the earth is at the center of the solar system and that the sun and planets orbit the earth. |
| **Geocentrism** | The theory that the sun is at the center of the solar system and that the earth and planets orbit the sun. |

2. Are your five senses always trustworthy? Why or why not?

______________________________

______________________________

______________________________

Hello, friend! One of the things I love about science is how it reminds us of truth from the Bible.

Me too! This week, we saw that our five senses can sometimes trick or deceive us. But that's not the only thing that can deceive us! Did you know that our hearts can also be deceitful? In Jeremiah 17:9, it says,

*The heart is deceitful above all things and beyond cure. Who can understand it?*

We often read about our heart in the Bible.

Is it talking about our actual heart? The one that beats in our chest and pumps blood through our bodies?

Great question! No, when the Bible talks about our heart, it is talking about who we are as a person — our minds, will, and even our emotions. Our minds, will, and emotions can be deceitful.

Like that time I took something that didn't belong to me. When Dad asked me what I had done, I felt like it would be better to lie about it. My heart was deceitful — the right choice was to tell the truth.

Yes. Our hearts can be deceitful. That is why it is important to know what God's Word says and choose to follow it no matter how we are feeling. This is how we renew our minds, like Romans 12:2 tells us:

*Do not conform to the pattern of this world, but be transformed by the renewing of your mind. Then you will be able to test and approve what God's will is—his good, pleasing and perfect will.*

I remembered that in the 10 Commandments, God told us not to lie. So, even though it was hard and my heart didn't want to, I chose to tell the truth. I'm glad I spend time each day renewing my mind by learning God's Word!

I have an idea, friend! Let's talk to Mom and Dad tonight during dinner about ways we can study God's Word together and renew our minds.

Look up Romans 12:2 in your Bible. You can ask your teacher for help if you need to. If you'd like, you can highlight this verse in your Bible. Memorize Romans 12:2 with your teacher or with a sibling. Then talk with your family about ways you can study God's Word together.

It sure was fun to talk about some science history this week. I'm looking forward to learning all about the scientific method next week!

Me too — as long as we don't have to taste vanilla again. Yuck!

I agree! Today it's time to add a new page to our Science Notebook. I think we should add a picture of vanilla extract so that we can remember our activity from this week. Vanilla extract is made from the vanilla bean. Vanilla beans grow on flowering vanilla orchid plants. I have a picture of a vanilla flower, beans, and vanilla extract that we can use as an example for our drawings!

Ooh, I like that idea! Let's get started on our Notebooks — we'll show you what ours look like. Be sure to have fun creating your Notebook picture!

In your Science Notebook, write:
**I learned that vanilla extract smells nice, but it tastes bitter!**

Then draw a picture of vanilla extract and the vanilla flower.

Our adventure this week reminded us that our hearts can be deceptive, just like our senses were. Romans 12:2 tells us to renew our minds so that we will know the will of God rather than be deceived by our hearts. Copy Romans 12:2 on the back of your Notebook page as a reminder.

*Do not conform to the pattern of this world, but be transformed by the renewing of your mind. Then you will be able to test and approve what God's will is—his good, pleasing and perfect will* (Romans 12:2).

# The Scientific Method

Day 1

Mm, mmm! Mom sure made us a good breakfast of pancakes, syrup, and bacon this morning!

I thought so too — and it will give us plenty of energy for today's adventure! Where should we begin?

Bacon.

Ben, we've already eaten the bacon that Mom made. I think it's time for you to take your mind off of bacon.

No, no, no, not that bacon! I'm talking about a different Bacon, Francis Bacon!

Oh Ben, you're always trying to trick me! Who was Francis Bacon?

**Francis Bacon**

Well, Francis Bacon was born in England in the year 1561. He worked in government and was known as someone who thought carefully about difficult questions in order to find answers or develop theories. Though he wasn't a scientist, Bacon made an important contribution to the field of science.

Francis Bacon recognized that the study of science was often disorganized and that people's senses and beliefs could lead them to incorrect conclusions. Francis Bacon argued that experiments were an important part of science. Experiments are activities that help us test our ideas and gain information about what we are studying.

Francis Bacon believed that scientists could use experiments to help them gain information. Scientists could then use that information to help them to reach the correct conclusions through logic. Through his work, Francis Bacon is recognized as the man who developed what we call the scientific method.

- [x] Baking soda
- [ ] Vinegar
- [ ] Lemon juice
- [ ] 2 bowls
- [ ] Tablespoon

**Weekly materials list**

Name: ______________________________

Ah, I know about the scientific method! The scientific method gives scientists a way to study and explore creation in an organized way. It helps us observe, question, and test things so that we can come to more accurate conclusions.

Yep! The scientific method guides us through a series of steps so that we can learn about God's creation. It also helps us avoid some of the problems in science that we talked about last week.

We'll be learning more about the scientific method this week. Before we end for today, though, I thought this quote from Francis Bacon was interesting. A quote is something that someone once said or wrote. Francis Bacon believed what is written in the Bible, and he said,

*There are two books laid before us to study, to prevent our falling into error; first, the volume of the Scriptures, which reveal the will of God; then the volume of the Creatures, which express His power (Men of Science, Men of God).*

1. What did Francis Bacon contribute to the study of science?

______________________________

______________________________

______________________________

______________________________

______________________________

2. Copy Francis Bacon's quote below. Are there any words you do not understand? Circle them, then ask your teacher to help you learn what they mean.

*There are two books laid before us to study, to prevent our falling into error; first, the volume of the Scriptures, which reveal the will of God; then the volume of the Creatures, which express His power.*

______________________________

______________________________

______________________________

______________________________

Today is the day we're going to learn all about the scientific method, so let's get started! The scientific method helps us to study science in an organized way as we ask questions, test our ideas, and develop conclusions. There are basically five steps to the scientific method — and they're easy to remember!

The **first step** of the scientific method is to make an observation. In other words, we notice something in God's creation that makes us curious.

Hmm, well, today I noticed that some flowers are growing really well in Mom's garden, but others seem to be withering and dying. They are all the same type of flower, though, so it made me curious.

That is a good observation! The **second step** of the scientific method is to ask questions about what we've observed. A few questions we might ask about the flowers would be: Why are those flowers withering and dying? What is different about those flowers? Should those flowers be withering, or is something wrong?

Once we've made an observation and asked a question about our observation, it's time for the **third step**! We need to create our **hypothesis** (said this way: hī-pŏth-ŭh-sĭs). A hypothesis explains what may be happening or gives a possible answer for the questions we've asked. A hypothesis is something that we can test through an experiment.

Ah, okay! So, my hypothesis is that these flowers need plenty of sunlight. The flowers that are dying are in the shade — I think they are not getting enough sunlight to stay alive.

Perfect! Do you think that if the flowers were moved into the sun they would grow well again?

I do!

It's time for **step number four** then! We need to test our hypothesis; this is also called an experiment. Experiments help us prove or disprove our hypothesis. Experiments give us information, or data, that can help us learn more.

Okay! In order to test my hypothesis, we'll need to ask Mom to help us dig up and move those flowers. Then, we'll need to watch for several days to see what happens.

Name: ____________________

That sounds like a good plan! And that brings us to the last step, **step number five**. Once we've done our experiment, we need to gather the results to share. We can take pictures or write down what happened — we'll talk more about that soon!

| |
|---|
| 1 Observation |
| 2 Ask Questions |
| 3 Hypothesis |
| 4 Experiment |
| 5 Results |

Match each step of the scientific method to the correct order. You can look back in the lesson to review if you need to!

| | |
|---|---|
| **Step 1** | Gather the results to share. |
| **Step 2** | Make an observation. |
| **Step 3** | Create a hypothesis. |
| **Step 4** | Test our hypothesis. |
| **Step 5** | Ask questions about our observation. |

Hello again, friend! Ben and I helped Mom move some of her flowers to the sunlight yesterday. But we're going to have to wait several days before we'll know if our hypothesis was correct.

In the meantime, I thought it would be fun to use the scientific method together in an experiment. We'll be able to see the results from this experiment right away!

**materials needed**

- Baking soda
- Vinegar
- Lemon juice
- 2 bowls
- Tablespoon

Ooh, this is going to be fun! Where should we start?

Well, do you remember the activity in *Adventures in the Physical World* when we used vinegar and baking soda to inflate balloons?

I sure do. That was one of my favorite activities!

Mine too! I've been thinking, vinegar is acidic, and it reacts with baking soda. I was helping Mom juice a lemon this morning, and she said lemon juice is also acidic. I wonder, do you think lemon juice will react with baking soda like vinegar does?

Ooh, it sounds like you've made an observation and asked a question!

I sure have! I observed that lemon juice is acidic, like vinegar. My question is, will lemon juice react with baking soda like vinegar does? Here is my hypothesis: Lemon juice is acidic like vinegar, and it will react with baking soda.

Let's test it!

## Activity directions:

1. Write your answer for question 1 on the worksheet on the next page.
2. Add 1 tablespoon of baking soda to each bowl.
3. Carefully pour the vinegar into the tablespoon, then pour it into one bowl. Observe what happens.
4. Rinse the tablespoon.
5. Carefully pour the lemon juice into the tablespoon, then pour it into the second bowl.
6. Observe what happens. Write your answers to questions 2–3 on the worksheet.

ASK PARENT FOR HELP

Name: ____________________

apply it

1. What do you think — will lemon juice react with baking soda like vinegar does? Your answer is your hypothesis. Write your hypothesis below.

2. Did the lemon juice and baking soda react like the vinegar and baking soda did?

3. Did your experiment prove or disprove your hypothesis?

I'm glad we learned about the scientific method this week! I'm looking forward to learning more about how it is used in science as we continue our adventure.

I've been thinking about something over the last two weeks, though, and I have a question. We've talked about a few of the ways science has been wrong in the past, like with the earth being at the center of the solar system. I thought that science was supposed to tell us the truth?

Ah, that is a really great question, Ben. Dad and I talked about this recently. Do you remember our definition of science?

Sure, science is the pursuit of knowledge and understanding about God's creation through an organized process. Science helps us to ask questions, test our ideas, and share what we've learned with others. Through science, we also learn more about God and our relationship with Him.

Right, science is the pursuit of knowledge and understanding. Remember, pursue means to chase after, seek, or search for something.

Through science, we seek to understand God's creation. But we also know that God's wisdom and knowledge are far beyond what we can fully understand. He is **omnipotent** (said this way: ŏm-nĭp-ŭh-tĕnt), which is a big word that means all mighty, with no limit. He is also **omniscient** (said this way: ŏm-nĭsh-ŭhnt), which is another big word meaning that He knows everything.

As humans, we are not all-knowing. We are also limited in our understanding and wisdom. As we study science, we pursue knowledge and understanding about God's creation. Sometimes, though, our understanding of science changes because we learn more or discover something new about God's creation. This is the pursuit of knowledge and understanding in action!

Science can help us understand more about God's creation, but it cannot give us complete truth; only God and His Word can do that. Though science may change at times, we know that God and the truth He reveals in Scripture never, ever change. James 1:17 says,

*Every good and perfect gift is from above, coming down from the Father of the heavenly lights, who does not change like shifting shadows.*

I'm glad God and His Word never change!

Look up up James 1:17 in your Bible. You can ask your teacher for help if you need to. If you'd like, you can highlight this verse in your Bible. Memorize James 1:17 with your teacher or with a sibling.

Woohoo! It's my favorite day of the week, the day we add another page to our Science Notebook! What should we add this week?

Hmm, we learned about the scientific method this week — let's add the steps to our Notebook! Since we don't have an example image to give us an idea this time, we'll need to use our own creativity! Let's write the steps of the scientific method and then draw something that will help us remember the steps.

Ah, so we could draw one picture of our experiment, or draw a different picture for each step of the scientific method.

Exactly, and remember to have fun as you create your Science Notebook! One thing I love about each of our Science Notebooks is that they are different. They remind me of God's creativity. We see God's creativity in creation, in the many different kinds of plants and animals He made. God gave us the ability to be creative just like Him. Each of our pictures are unique and display the creativity God has given us.

Let's get started! Hannah and I are going to draw a picture to illustrate each step of the scientific method. Sam drew a picture of an experiment he did in the past. We can't wait to see your Notebook!

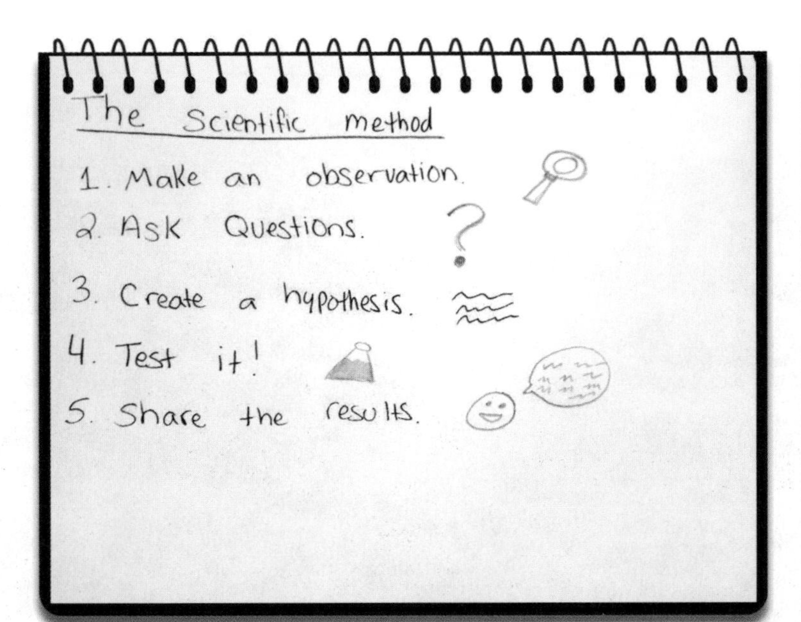


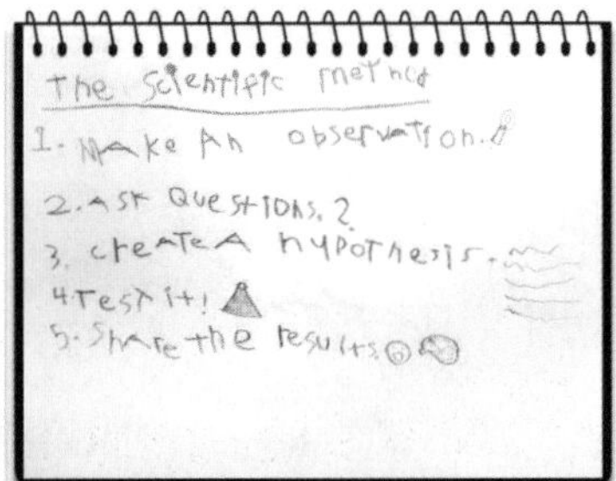

In your notebook, write the steps of the scientific method:

1. Make an observation.
2. Ask questions.
3. Create a hypothesis.
4. Test it!
5. Share the results.

Then draw a picture of your experiment, or draw small pictures to help you remember each step.

Our science adventure this week reminded us that though science may change at times, God and His Word never change. We can trust what we read in God's Word. Copy James 1:17 on the back on your Notebook page as a reminder.

*Every good and perfect gift is from above, coming down from the Father of the heavenly lights, who does not change like shifting shadows* (James 1:17).

week 4

# Lab Reports

Day 1

Welcome back for another science adventure! We're going to continue exploring the scientific method this week — and our mom has a fun activity planned! But first, let's review the scientific method. Can you remember the five steps? Here they are:

1. Make an observation.
2. Ask questions.
3. Create a hypothesis.
4. Test it!
5. Share the results.

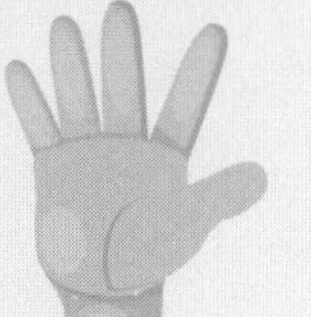

That reminds me, Hannah, how are Mom's flowers? Are they growing any better?

Oh, I almost forgot to give you an update on those! Mom and I moved the flowers to a sunny area last week, and we've started to see a change — the flowers are starting to grow really well again!

Yay, our hypothesis was correct! I'm glad you were able to help save Mom's flowers.

Me too! I've been thinking about how we can document our experiments and results, though. We're going to have many different experiments as we study science this year, and it will be easy to forget about the things we've tested and learned.

That's a good observation, Hannah! Documenting our experiments and their results is actually part of how we work through the scientific method. As we work through the five steps, we'll need to record the information, or data, we gather along the way.

One way we can record data that we've gathered is through a lab report. Lab reports help us remember the experiments we've done, what we learned from them, and can even help us ask new questions. A lab report gives us a way to write down our question, hypothesis, how we tested it, and the results of our experiment.

We'll be talking more about lab reports this week. But first, let's record what happened with our mom's flowers in a comic strip!

**Weekly materials list**

- 2 ballons ✓
- Vinegar
- 2 soda bottles (same size)
- Lemon juice
- Baking soda
- Funnel
- Soft tailor's tape measure
- Permanent marker
- Tablespoon
- ½ measuring cup

Name: ____________________

apply it

1. Hannah made an observation, asked a question, created a hypothesis, tested it, and helped save Mom's flowers. Use your imagination to draw a comic strip showing what happened.

| | | |
|---|---|---|
| | | |
| Mom's flowers were dying. | Hannah and Mom moved the flowers into the sunshine. | The flowers are growing well again! |

2. Have you made an observation and asked a question before? What was your question?

Welcome back, friend! Today, we're going to learn how to document our science experiments in a lab report.

I'm excited to learn about lab reports because sharing what I've learned with others is one of my favorite parts of science! Creating lab reports will help me make sure I don't forget important or interesting details.

Let's dive right in, then! A lab report gives us an organized way to record and share our experiments. It can be very simple or quite complex — but either way, the goal is to record data by writing things down or by taking pictures.

Dad helped me make an example lab report last night. Let's study it together!

The first thing I notice is that we need to write our name and the date at the top. This will tell others who did the experiment and when the experiment was done.

Exactly. Next, we need to write down our question. For this lab report, Dad and I used Mom's flowers as an example. The question was, "Why are some of Mom's flowers withering and dying?"

After that, it's time to write down our hypothesis. Hannah's hypothesis was, "The flowers need plenty of sunshine. If we move the flowers to a sunny area, they will grow well again."

Then we tested it! Mom and I moved the flowers, and then we waited to see what would happen.

We can write down anything we notice during our experiment in the "Things I observed" section. This can help us remember anything interesting or curious we see, and it may even help us ask new questions!

I noticed that Mom's flowers started growing much better again three days after we moved them, so Ben wrote that in "Things I observed."

Finally, we need to write down the result of our experiment. Our hypothesis was correct; once we moved Mom's flowers, they began to grow well again!

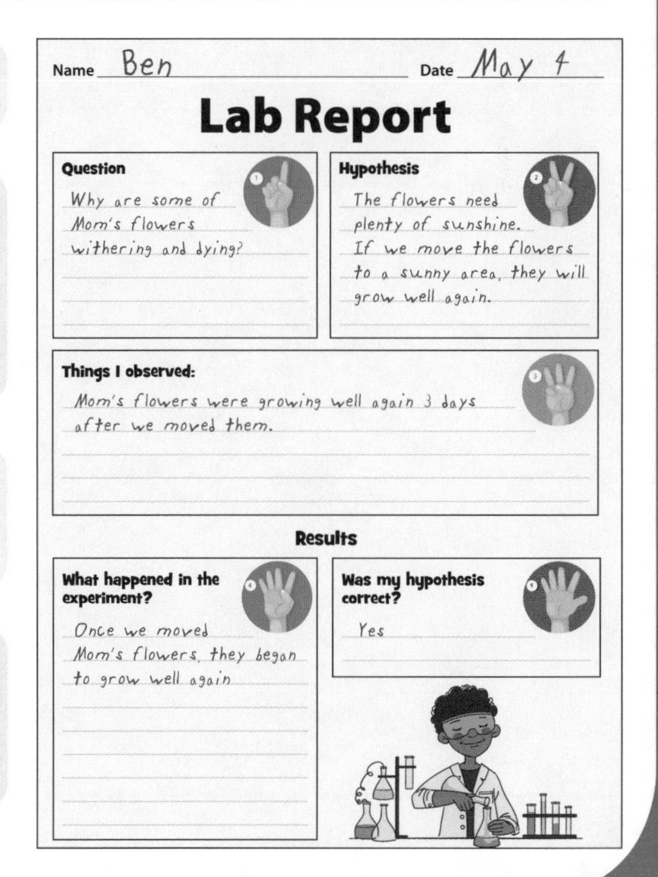
Name Ben Date May 4

Lab Report

Question
Why are some of Mom's flowers withering and dying?

Hypothesis
The flowers need plenty of sunshine. If we move the flowers to a sunny area, they will grow well again.

Things I observed:
Mom's flowers were growing well again 3 days after we moved them.

Results

What happened in the experiment?
Once we moved Mom's flowers, they began to grow well again

Was my hypothesis correct?
Yes

Name: ______________________________

apply it

1. What does a lab report help us do?

2. How can a lab report help us as we continue to explore science?

Name ______________________ Date ______________

# Lab Report

### Question

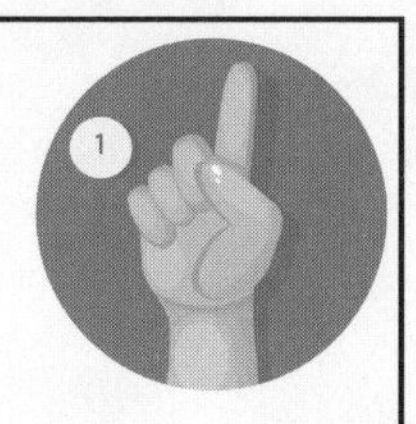


### Hypothesis

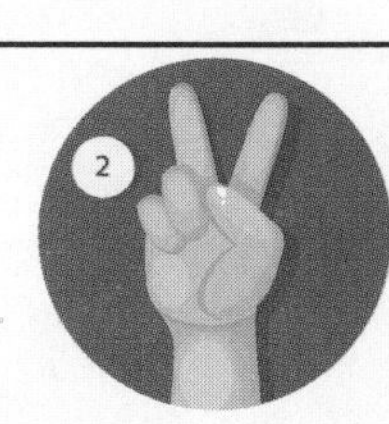


### Things I observed:

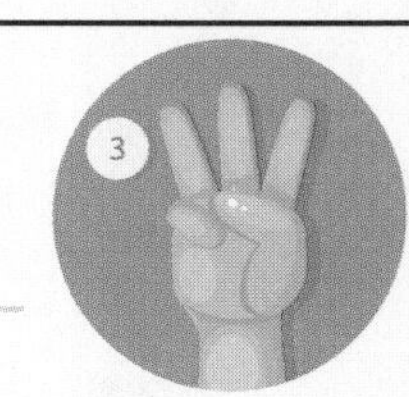


## Results

### What happened in the experiment?

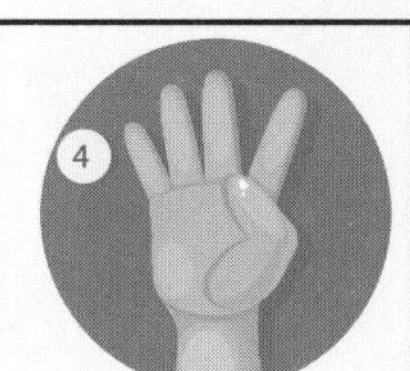


### Was my hypothesis correct?

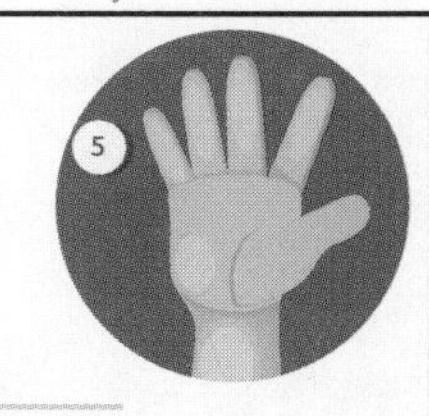

# Additional Lab Notes

Oh good, you're here! We're going to practice creating a lab report today — and I have the perfect idea for an experiment! Last week, we did an experiment to see if lemon juice reacts with baking soda like vinegar does. We discovered that both lemon juice and vinegar react with baking soda. Our experiment gave me a new question: does one react more than the other?

**materials needed**

- 2 balloons
- 2 soda bottles (same size)
- Vinegar
- Lemon juice
- Baking soda
- Funnel
- Soft tailor's tape measure
- Permanent marker
- Tablespoon
- ½ measuring cup

Hmm, so your question is, does lemon juice or vinegar have a stronger reaction with baking soda?

Right! Since the reaction between baking soda and lemon juice or vinegar creates gas bubbles, we can use the reaction to blow up balloons — just like we did in *Adventures in the Physical World*. Then we can measure the reactions by measuring how full each balloon becomes!

Okay, let's get started! I have a lab report right here. We can fill it out as we do our experiment. Let's write our name and the date first, then we can write our question and hypothesis! Don't forget to ask your teacher to help you with this experiment before you get started.

## Activity directions:

1. Write your name and the date on the lab report on the previous page.
2. Write your question. An example from the lesson is, "Does lemon juice or vinegar have a stronger reaction with baking soda?"
3. Now write your hypothesis. How do you think vinegar and lemon juice will react — will one react more than the other? Hannah and Ben's hypothesis was, "Since lemon juice and vinegar are both acids, we think they will react the same with the baking soda."

4. Let's test it now! Ask your teacher to help you use the funnel to add 2 tablespoons of baking soda to each balloon. Rinse out the funnel.

5. Use the funnel to pour ½ cup of vinegar into one soda bottle. Use the permanent marker to write 1 on this bottle. Rinse out the funnel, then pour ½ cup of lemon juice into the other soda bottle. Use the permanent marker to write 2 on this bottle.

6. Hold the first balloon off to the side (so that the baking soda doesn't pour into the bottle yet) and carefully stretch the mouth of the balloon over the opening of the soda bottle. You can ask your teacher for help if you need to. Repeat with the second soda bottle.

7. Now pick up the first balloon to dump the baking soda inside bottle number 1. Observe what happens. Once the balloon has finished inflating, use the soft measuring tape to measure around the widest part of the balloon. Write "Bottle number 1, baking soda and vinegar" in the "Things I observed" section of the Lab Report, then write down the measurement.

8. Repeat that process now with bottle number 2. Write "Bottle number 2, baking soda and lemon juice" in the "Things I observed" section of the Lab Report, then write down the measurement.

9. Now it's time to fill out the Results section. Was your hypothesis correct? Did the balloons blow up to the same amount, or was one bigger than the other? If one was bigger, which bottle had the stronger reaction?

I'm so glad we've learned how to use lab reports to document the information we collect through our experiments. Our lab reports will help me remember all the experiments we're going to do this year — they're such a great reminder of what we've learned!

Learning about lab reports this week also reminded me of something I learned in Sunday school a few weeks ago.

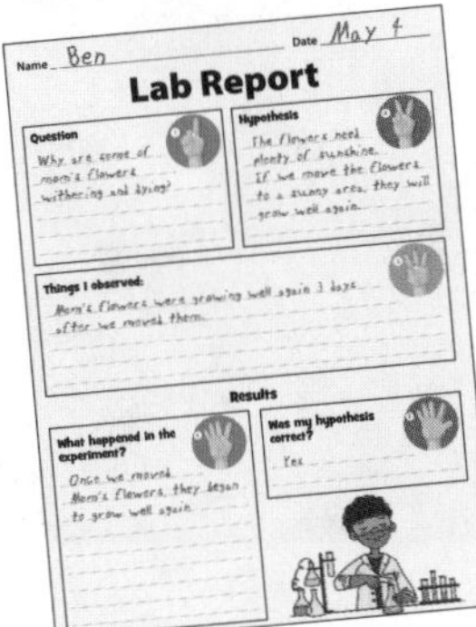
Name Ben Date May 4

**Lab Report**

**Question**
Why are some of mom's flowers withering and dying?

**Hypothesis**
The flowers need plenty of sunshine. If we move the flowers to a sunny area, they will grow well again.

**Things I observed:**
Mom's flowers were growing well again 3 days after we moved them.

**Results**

**What happened in the experiment?**
Once we moved Mom's flowers, they began to grow well again.

**Was my hypothesis correct?**
Yes

Oh really? What would that be?

Well, we talked about different times in the Bible when God told His people to create a monument of stones or to celebrate a feast to remind them of God's faithfulness.

One of those stories is in the book of Joshua. In Joshua, we read that it was time for the nation of Israel to enter the land God had promised them — but they needed to cross the Jordan River first. God stopped the flow of the Jordan River so that the nation could cross on dry ground — just like God did when they left Egypt and crossed the Red Sea. Let's read what happened after they crossed the Jordan River in Joshua 4:1–3,

*When the whole nation had finished crossing the Jordan, the LORD said to Joshua, "Choose twelve men from among the people, one from each tribe, and tell them to take up twelve stones from the middle of the Jordan, from right where the priests are standing, and carry them over with you and put them down at the place where you stay tonight."*

Hmm, why did God want them to carry those stones?

The answer to that is found in Joshua 4:6b–7:

*In the future, when your children ask you, 'What do these stones mean?' tell them that the flow of the Jordan was cut off before the ark of the covenant of the LORD. When it crossed the Jordan, the waters of the Jordan were cut off. These stones are to be a memorial to the people of Israel forever.*

The Jordan River

Ah, so those 12 stones would be a reminder to the people of when God stopped the flow of the Jordan River so that His people could cross over.

We see throughout the Bible, and even in our own lives, that it is easy to forget God's faithfulness. God gave His people reminders of what He had done in their lives through memorials, like those 12 stones. Whenever someone saw those stones, it would have reminded them of the time God stopped the flow of the Jordan. It would have reminded them of God's faithfulness and power.

We see God's work and faithfulness in our own lives too. We can use reminders to help us remember all the times God has been faithful to us.

That's why Mom has special Bible verse pictures on our walls! She told me they remind her of God's faithfulness.

Yes! Mom and Dad also write down things God has done in our lives so that we can read about them later. That gives me an idea — let's begin our own journals so that we can write down the things we've learned and the ways God has been faithful in our lives. It will be fun to read what we've written down and remember how God has been faithful to us later on. You can use a piece of paper or a notebook, friend, if you'd like to begin writing down what God has done in your life too!

Does your family have any reminders of things the Lord has done in your lives? Ask your parents, grandparents, or other family members if they can share any stories of God's faithfulness. Then look up Joshua 4:6–7 in your Bible. If you'd like, you can highlight these verses.

We've reached the end of another week, and it's time to add a new page to our Science Notebook!

I have my Notebook right here, and I'm ready to draw — I think I have a good idea for this week. Scientists often keep their lab reports in a lab notebook. This helps them keep all of their experiments together in one place. We can draw a picture of a lab notebook!

Ooh, I like that idea. I have an example image of a notebook right here. Once we've drawn the notebook, we can write Lab Notebook on its cover.

Let's get started on our drawings. We'll be sure to show you our Science Notebooks once they are done — and don't forget, have fun creating your picture!

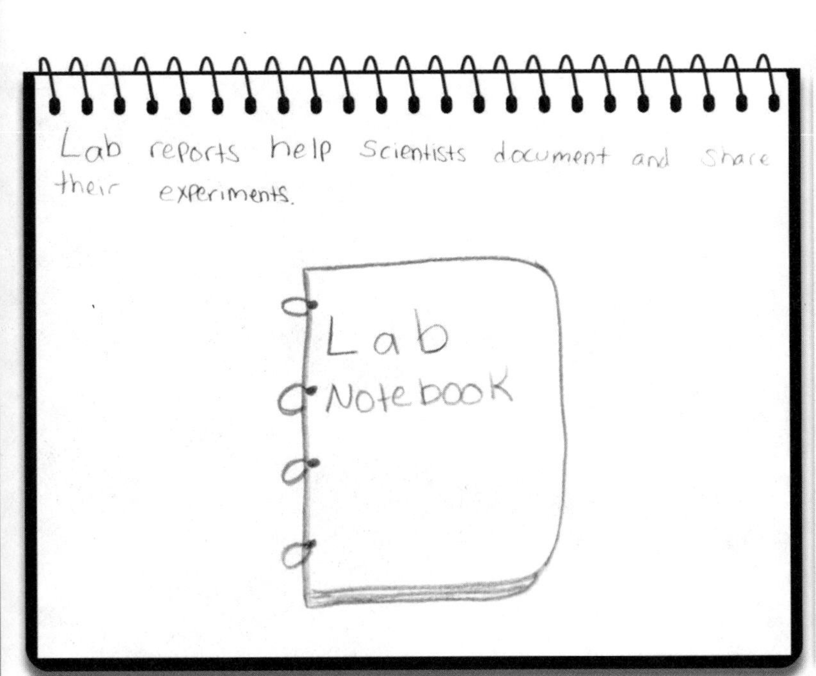


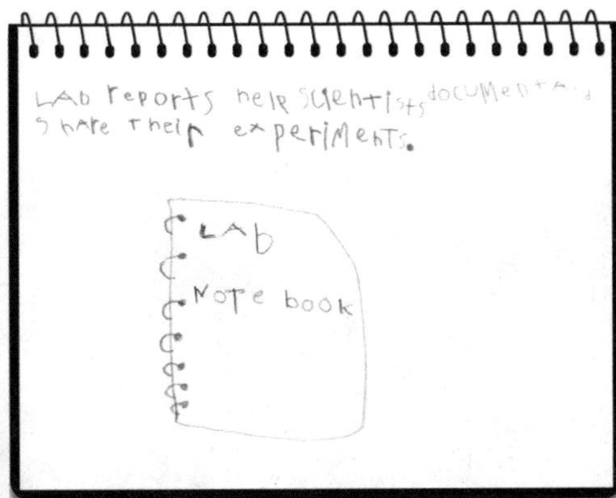


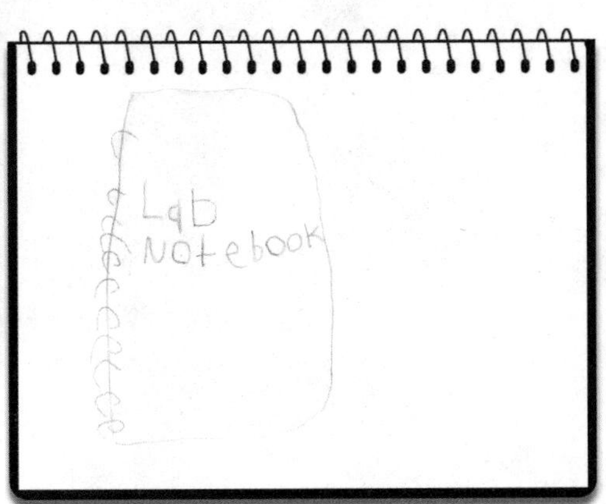

In your Notebook, write: Lab reports help scientists document and share their experiments.

Then draw a picture of a notebook and write "Lab Notebook" on the cover.

Learning how to record our experiments in a lab report this week reminded us about the ways God told His people to record and remember what He had done for them. Copy Joshua 4:7 on the back of your Notebook page as a reminder.

*. . . tell them that the flow of the Jordan was cut off before the ark of the covenant of the* L*ORD*. *When it crossed the Jordan, the waters of the Jordan were cut off. These stones are to be a memorial to the people of Israel forever* (Joshua 4:7).

The Jordan River

week 5

# Day 1 Systems of Measurement

We're back and ready for another science adventure together! So far, we've developed our definition of science, learned some of the problems scientists faced in the past, discovered how the scientific method brought an organized structure to science, and practiced creating our own lab reports to share information.

We've explored a lot in a short amount of time already! But our adventure is definitely not over yet. We have so much more to learn! This week, we're going to talk about measurements.

Measurements help us communicate the length, weight, or amount of something. Do you remember the measurements from our experiment with baking soda and vinegar? The instructions told us to use ½ cup of vinegar and 2 tablespoons of baking soda. A cup and a tablespoon are both ways we use to measure things.

Right! Measurements are an important tool we use in science. If we didn't have a way to measure things, we wouldn't be able to tell someone else how to repeat an experiment we've done.

Imagine if the directions for our vinegar and baking soda experiment didn't have any measurements. We wouldn't have known how much vinegar or baking soda to use. We would have had to guess — and if we guessed wrong, it could have caused our experiment to fail.

Measurements are important in science and in our everyday lives. We use measurements in school, when we go shopping, and even in the kitchen. Last week, I used measurements to help me make biscuits for dinner. But I accidentally mixed up the teaspoon and the tablespoon, and I added way too much baking soda to the biscuits. They tasted awful! It was a good reminder to carefully follow directions — and make sure you are using the right measurement.

- Measuring tape ✓
- Bathroom scale
- Gallon container
- Ruler

**Weekly materials list**

Name: ____________________

That's for sure. We're going to be learning about the two systems we can use to help us measure things this week, and I can hardly wait!

1. Have you followed directions or measurements to help you make something?

2. If you have, what did you make?

3. Have you ever made a mistake measuring something, like Hannah did?

4. If so, what happened?

**materials needed**

- Measuring tape
- Bathroom scale
- Gallon container

We're exploring the two systems we can use to measure things this week! Let's begin with the **imperial** (said this way: ĭm-pēr-ē-ŭhl) system. The imperial system can also be called the English system, and it is the system most commonly used to measure things in the United States.

A system of measurement must have a standard and consistent way to measure things. We call these standards units of measurement. The imperial system uses units called inches, feet, ounces, pounds, miles, and gallons to measure or weigh objects. You may have even used these units yourself!

Though these units of measurement had been used for many hundreds of years, the imperial system did not become an official system of measurement until the year 1824, when Great Britain declared it the official measurement standard.

During that time, Great Britain had authority over many different colonies around the globe, and it was known as the British Empire. Imperial is a word that refers to an empire or emperor. Since this system of measurement became the official way to measure things in the British Empire, it was called the imperial system.

Measurements must be clear, consistent, and accurate. Since there are different systems and units of measurement that someone could be using, it is important that we know which system and which unit a person used to measure an object.

For example, when I helped Dad install the new floor in our house, he needed to cut a board. It wouldn't have been helpful for me to tell him to cut the board to 3 — he would need to know whether to cut the board to 3 inches or 3 feet. There is a huge difference between those two units of measurement!

The imperial system uses abbreviations to show which unit of measurement has been used. An **abbreviation** (said this way: ŭh-brē-vē-ā-shŭhn) is a shorter way to write a word. You may also see a symbol used for inches and feet. These are some of the abbreviations and symbols for the imperial system:

| Measurement | Abbreviation | Example |
|---|---|---|
| Inches | in or " | 3 in or 3" |
| Feet | ft or ' | 3 ft or 3' |
| Ounces | oz | 4 oz |
| Pounds | lb | 8 lb |
| Gallons | gal | 1 gal |
| Miles | mi | 100 mi |

Name: ______________________________

When we write a measurement, we first write the amount of the measurement and then the unit that we used to measure. So, if my pencil measures 5 inches, I would write down 5 in. We'll talk more about the other standard system of measurement tomorrow — for today, let's use the imperial system to measure some objects!

A measuring tape can help us measure an object's length in inches. Find each object below in your house and measure it in inches. Write the measurement on the line — don't forget to write 'in' after the measurement so others will know that you measured in inches.

1. Spoon: ______________________________

2. Fork: ______________________________

3. Book: ______________________________

4. Pencil: ______________________________

5. Toy: ______________________________

6. A bathroom scale can tell us how many pounds we weigh. Weigh yourself using a bathroom scale. How many pounds do you weigh? Don't forget to write 'lb' after the measurement so that others will know you used pounds as your measurement unit.

______________________________

7. A gallon is a unit we use to measure liquids. Fill up the gallon container with water. Do you think a gallon is light or heavy?

______________________________

I'm glad you're back! It's time to learn about the second system of measurement today. This system of measurement is called the **metric** (said this way: mĕ-trĭk) system.

In our last lesson, Hannah mentioned that measurements must be clear, consistent, and accurate. Before Great Britain made the imperial system an official system of measurement, the measurements in this system actually weren't very standard or consistent.

**materials needed**

- Measuring tape
- Ruler

That's right! The units people used to measure things weren't always based on a standard. This meant that an inch or a pound could be different in different parts of the British Empire! That led to a lot of chaos and confusion.

Since the units weren't standard, it also meant that the imperial system was not a good system of measurement for scientists. The measurement a scientist made in one part of the world needed to be the exact same measurement for a scientist somewhere else in the world.

This led scientists in France to create a new system of measurement that would follow a clear, consistent standard all around the world. The metric system was designed to be a clear and consistent way to measure things. It uses multiples of 10, which makes it very easy for scientists to use. France began using the metric system in 1795, and it eventually became the official system of measurement for most of the world.

The basic units of measurement in the metric system are called meters, grams, and liters. Just like the imperial system, we can also use abbreviations in the metric system to tell someone which unit we used to measure.

| Measurement | Abbreviation | Example |
|---|---|---|
| Meters | m | 3 m |
| Grams | g | 3 g |
| Liters | L | 4 L |

A meter, gram, and liter are the basic units of measurement in the metric system — but sometimes we need a way to measure something that is smaller or larger than the basic unit. Rather than have completely different words, like the imperial system does, the metric system uses prefixes. A **prefix** (said this way: prē-fix) is a word placed in front of another word, like this: centimeter (said this way: sĕn- tŭh-mē-ter).

Ah, so meter is the basic unit and "centi" is the prefix! The ruler that I use for school can measure in centimeters — the abbreviation is "cm."

The prefix "centi" means 100. It would take 100 centimeters to equal one full meter. Centimeters help us measure small objects. Another prefix that we use in the metric system is "kilo" (said this way: kē-lōh). "Kilo" means 1,000, so 1 kilometer would equal 1,000 full meters.

Name:

The metric system uses different prefixes to help us measure things that are much smaller than the basic units of meter, gram, and liter, as well as things that are far larger. The metric system is now used around the world — and it is also the system of measurement that scientists use.

1. One meter is just over 39 inches long. Use the measuring tape to measure 39 inches. Is a meter shorter or longer than you expected?

2. Does your family use the imperial system or the metric system to measure things?

3. What do you think it would be like if measurements didn't follow a standard around the world?

4. Use the ruler to measure the width of your hand in inches and then in centimeters. You can ask your teacher for help if you need to. Write down the measurements on the lines.

   The length of my hand is:

   ______________ inches

   ______________ centimeters

I had fun learning about systems of measurements this week! Did you know that people in Bible times also had units of measurements? One unit of measurement was called a **cubit** (said this way: kyoo-bĭt). A cubit was the length from an adult's elbow to the tip of their middle finger. In the book of Genesis, God gave Noah instructions and cubit measurements for building the Ark. We can read the instructions in Genesis 6:14–15,

*So make yourself an ark of cypress wood; make rooms in it and coat it with pitch inside and out. This is how you are to build it: The ark is to be three hundred cubits long, fifty cubits wide and thirty cubits high.*

Using the imperial system of measurement today, that means the Ark was around 510 feet long, 85 feet wide, and 51 feet high!

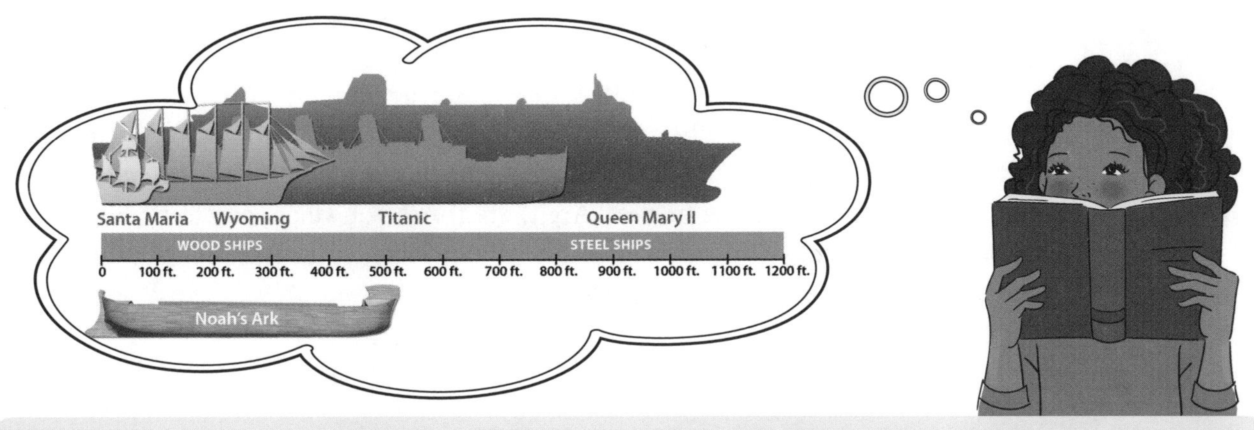


You may also see the word *ephah* (said this way: ē-fŭh) or *hin* in the Bible. These were also units of measurement. An *ephah* was a way to measure something dry, like wheat, and a *hin* was a way to measure liquids.

But whether we use measurements from the Bible, the metric system, or the imperial system, the Bible also tells us that it is important to God that our measurements are accurate and honest.

In the book of Leviticus, God gave the Israelites His law. He told them how they were to live set apart for Him. In Leviticus 19:35–36, God told them:

*Do not use dishonest standards when measuring length, weight or quantity. Use honest scales and honest weights, an honest ephah and an honest hin. I am the LORD your God, who brought you out of Egypt.*

If someone were to use a dishonest standard, they might make their measurement just a bit bigger or smaller than the actual measurement. A dishonest measurement cheats someone else. But God's Word tells us that we are to be honest when we deal with others. Tonight, let's talk with our friends or family about ways we can be honest with others!

There are times in our lives when we may be tempted to be dishonest or to cheat. Talk to your family about times they may have been tempted and ask them to tell you about a time they chose to be honest. Are there times you've chosen to be honest? Then look up Leviticus 19:35 in your Bible. If you'd like, you can highlight this verse.

I'm excited to add a new page to our Science Notebook today! It was really interesting to learn about the two systems of measurement — and I have a feeling we'll be talking more about the metric system next week.

I have an idea for our drawing this week. Let's use our ruler to draw a line that is five inches long. Then we can draw another line using centimeters.

Ooh, I like that idea! Let's draw a line that is five centimeters long. The lines will be different lengths to show the different units of measurement. Let's get started!

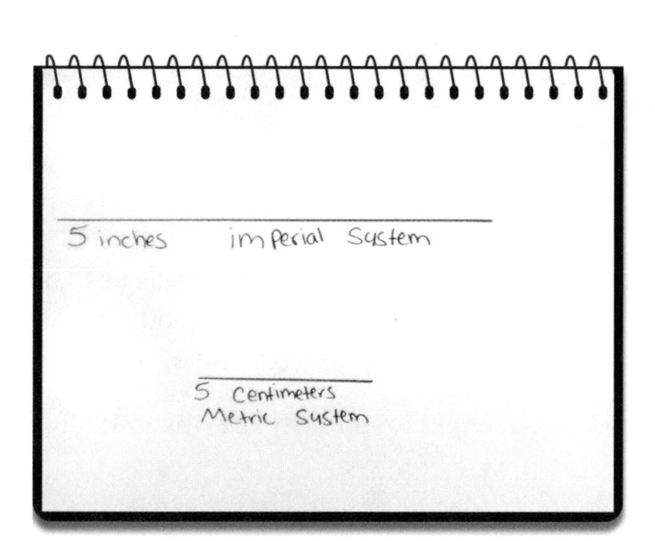


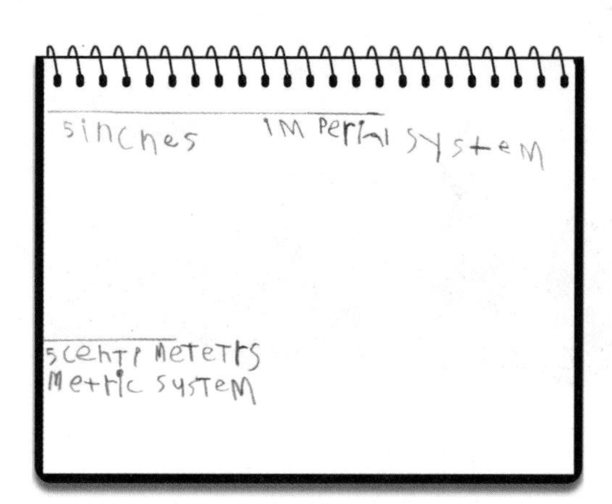

In your Notebook, use a ruler to draw a 5-inch line. Label this line "5 inches, imperial system."

Then draw a line that is 5 centimeters long. Label this line "5 centimeters, metric system."

Learning about measurements this week reminded us that God tells us to be accurate and honest in the way we measure things and deal with others. Copy Leviticus 19:35 on the back of your Notebook page as a reminder.

*Do not use dishonest standards when measuring length, weight or quantity* (Leviticus 19:35).

# Mass & Matter

Hello, friend, I'm glad you're back for another science adventure! Where should we begin today, Hannah?

Well, first we need to continue our discussion on measurements. Remember that in science, measurements are very important — and it's also important that our measurements are consistent.

Weight is one type of measurement that we use in our everyday lives. For instance, mom weighs the fruits and vegetables at the grocery store to make sure we buy enough pounds of each. We also weigh our bodies on the bathroom scale. We often measure the weight of something, but what exactly is weight anyway?

To answer that question, we need to start with gravity. Gravity is the force that pulls us toward the earth. No matter how high up you can jump, gravity pulls your body back down. Weight measures the force of gravity pulling down on an object.

Ah, but there is a problem with the measurement of gravity. The force of gravity isn't always consistent. For example, there is much less gravity on the moon than there is on the earth. If I weigh 60 pounds here on earth, I would only weigh around 10 pounds on the moon. That's because the force of gravity isn't as strong on the moon.

That is a big difference in weight! But you would still be the same person and the same size on the earth or on the moon. Your size and shape don't change — just the force of gravity.

Scientists need measurements that are clear and consistent — whether an object is here on earth or on the moon. So, scientists often use the measurement of mass instead of weight.

- Kitchen scale that measures in grams or kilograms ✓
- 2 balloons
- Yarn
- Permanent marker
- Scissors
- Ruler

**Weekly materials list**

Name: ______________________________

**Mass** (said this way: măs) measures the amount of material in something or someone, rather than the force of gravity. Because the amount of material in something or someone doesn't change with gravity, the measurement of mass stays consistent. We use grams and kilograms to measure the mass of an object.

Ah, so mass is a part of the metric system! I can't wait to continue our discussion tomorrow.

apply it

1. Why do you think it is important for measurements to stay consistent?

______________________________

______________________________

______________________________

______________________________

We can measure objects in grams or kilograms using a kitchen scale. Ask your teacher to help you set up the scale to measure in grams or kilograms. Pick three objects to measure from the kitchen. Then fill in what you measured and its measurement, and circle the unit of measurement you used.

2. Object 1 was ______________________________, and it weighed ______________________________

grams / kilograms

3. Object 2 was ______________________________, and it weighed ______________________________

grams / kilograms

4. Object 3 was ______________________________, and it weighed ______________________________

grams / kilograms

It sure was interesting to learn about mass yesterday. I also had fun measuring mass in the kitchen! Today, we're going to shift our focus and begin exploring the field of chemistry together. Where should we start, Hannah?

Hmm, I think we should start where it matters most — with matter!

Hey, I usually handle the jokes around here!

I know, but I couldn't resist. Ben, take a look around the room. What do you see?

Well, I see you, the table, my school books, a glass of water, a plate of cookies, the inside of our house, and some trees through the window.

Everything that you see is actually known as matter in chemistry. **Matter** (said this way: măt-er) is anything that takes up space and has mass.

Ooh, I remember learning about matter in *Adventures in the Physical World*. Let's review it quickly. Matter can be a liquid, a solid, or a gas. An example of matter in a liquid state is water. Liquids can be poured, and they fill a space. If we froze the water, we would create ice, which is matter in a solid state. In a solid state, matter takes up a space and has a defined shape. If we then melted the ice in a pot on the stove, the water would evaporate through steam — which would be matter in a gas state! When matter is in a gas state, it doesn't stay in a consistent shape unless it is filling a defined space, like a balloon.

Thanks, Ben! Matter is all around us from the things that we can see to the air we breathe. The field of chemistry explores matter, as well as how it reacts and responds. Understanding what matter is and how it responds helps us to better explore the living and non-living things around us. Chemistry gives us a foundation to study many other branches of science.

Name:

apply it

1. Copy the definition of matter: Matter is anything that takes up space and has mass.

2. What matter do you see in the room around you?

3. Draw a picture of matter you see around you.

I sure am having fun learning about mass and matter this week! Let's review what we've learned so far. Mass measures the amount of material in something or someone. Mass is part of the metric system, and we can use grams or kilograms as our unit of measurement.

Matter is anything that takes up space and has mass. Everything we can see — and even the things we can't see, like air — is matter. Matter can be a solid, liquid, or gas. I understand solids and liquids being matter because it is easy to see that they take up space and have mass. But it's harder for me to understand how something like gas can also be matter.

**materials needed**

- [ ] 2 balloons
- [ ] Yarn
- [ ] Permanent marker
- [ ] Scissors
- [ ] Ruler

Ah, I know what you're saying. A gas could be steam rising from a pot, the oxygen gas we inhale, or the carbon dioxide gas we exhale. Since we often can't see a gas, it's harder to observe how it takes up space and has mass.

I have an activity we can do that will help us understand how gas is matter — let's get right to it!

## Activity directions:

ASK PARENT FOR HELP

1. Write your name and the date on your lab report (on the next page). Next, write the question: Does gas take up space and have mass?
2. Blow up the first balloon as big as you can — but be careful not to pop it! Tie the balloon and use the permanent marker to write "1" on this balloon.
3. Use just 2–3 breaths to blow up the second balloon. Tie the balloon and use the permanent marker to write "2" on this balloon.
4. Measure and cut two 12-inch pieces of yarn. Be very careful to cut both pieces of yarn to the same size. Next, take the first piece of yarn and carefully tie it to the first balloon. Tie the other end around one edge of the ruler. Repeat with the second balloon and tie it to the other edge of the ruler.
5. Slide each string to the edge of the ruler.
6. We're going to use the ruler like a scale. Do you think our scale will show the balloons are equal, or will one side of the ruler lower to show that one balloon has more mass than the other? Write your answer in the Hypothesis section of your lab report.
7. Pick up the ruler now and gently hold it in the middle between your thumb and index finger. Be sure to hold it in the middle very gently. Does the ruler remain balanced or does it tip to the side of one balloon or the other? Once you've finished observing, finish your lab report to record what you observed.

Name ______________________ Date ______________

# Lab Report

## Question

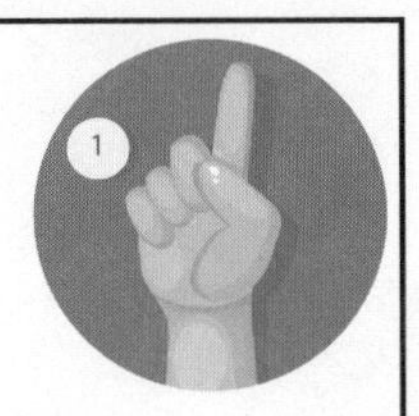

## Hypothesis

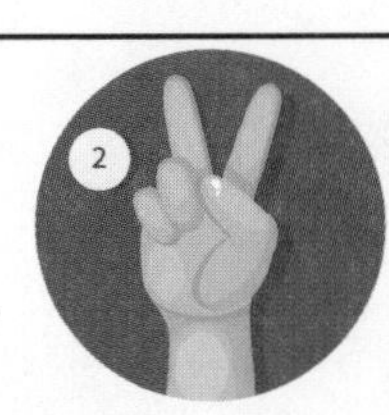

## Things I observed:

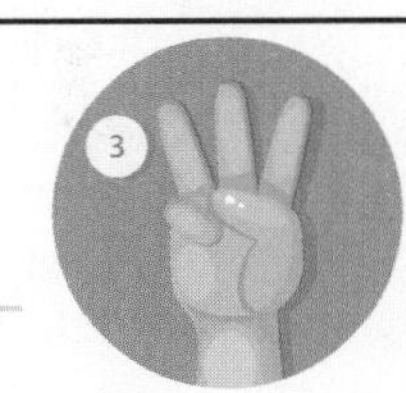

## Results

### What happened in the experiment?

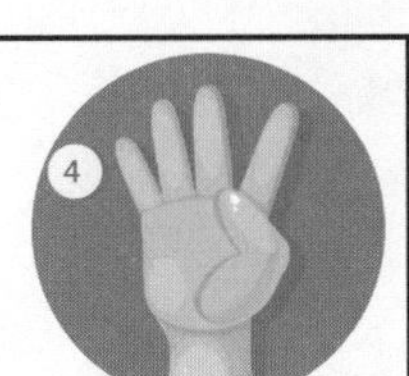

### Was my hypothesis correct?

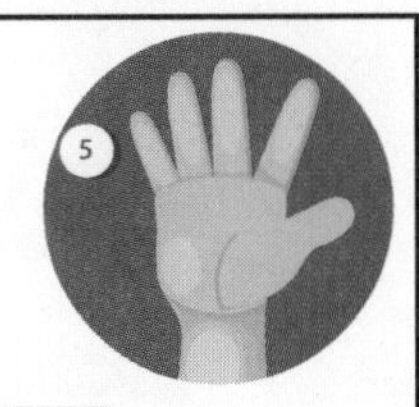

# Additional Lab Notes

Hasn't it been fun to learn about mass and matter this week? I'm excited to continue our exploration of chemistry together over the next few weeks!

Me too! As we've been learning this week, I've been thinking about how scientists need measurements to remain consistent — no matter where they are. Just like scientists need a consistent standard to measure things by, we also need a consistent standard of what truth is. What we believe to be truth, as well as right and wrong, will be the standard that we live our lives by.

This week, Mom and Dad were talking to us about how we as humans often change our standards for what is truth, right, and wrong. We need a consistent standard that doesn't change to live our lives by.

And we find that in the Bible!

Yes! God's standard of what is truth, right, and wrong does not change. God revealed His standard to us in His Word, the Bible. In Numbers 23:19 we read,

*God is not human, that he should lie, not a human being, that he should change his mind. Does he speak and then not act? Does he promise and not fulfill?*

Dad told us that we will see the standards for truth, right, and wrong change in the world around us as we live our lives. But no matter what the world around us says, God's truth and standards in the Bible never change, and they are the only consistent truth we can live our lives by.

I'm very glad that God's standard doesn't change. His Word is truth, and it gives us a firm foundation to build our lives upon. Though it's not always easy to live our lives by God's standards, He promises to be with us, and James 1:12 tells us,

*Blessed is the one who perseveres under trial because, having stood the test, that person will receive the crown of life that the Lord has promised to those who love him.*

What a good reminder; thanks for sharing with us today, Hannah!

Talk to your family about how you're building your life on God's standard. Then look up Numbers 23:19 in your Bible and memorize it with your teacher or with a sibling. If you'd like, you can highlight this verse in your Bible.

Whew, what a week we've had exploring mass and matter! Are you ready to add a new page to our Science Notebook?

I sure am! Let's draw a picture of matter in three states: solid, liquid, and gas. I have a picture right here that we can use for an example.

Ooh, I like it! Remember to have fun creating your Science Notebook.

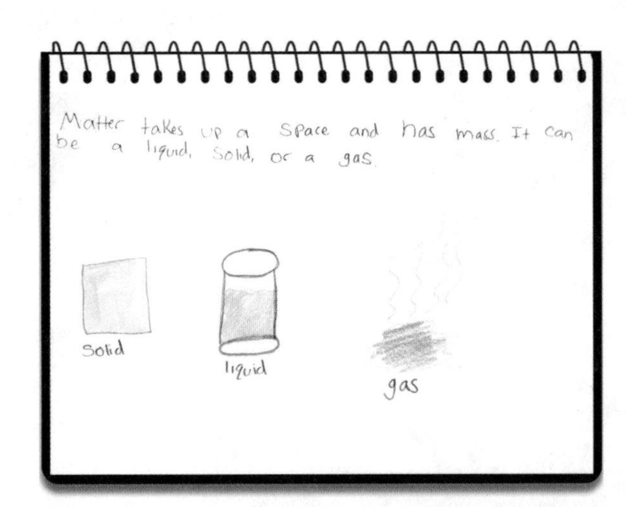

In your Notebook, write: Matter takes up space and has mass. It can be a liquid, a solid, or a gas.

Then draw a picture of matter in a solid, liquid, and gas state.

Learning about mass and matter this week reminded us that God's Word is consistent. God's standard and the truth revealed in the Bible does not change. Copy Numbers 23:19 on the back of your Notebook page as a reminder.

*God is not human, that he should lie, not a human being, that he should change his mind. Does he speak and then not act? Does he promise and not fulfill?* (Numbers 23:19).

# Day Describing Matter

Welcome back for another science adventure! We learned about mass and matter in our last adventure. Ben, can you remind us what mass and matter are?

I sure can! Mass is the way we measure the amount of material in someone or something. We can measure mass in grams or kilograms. Matter is anything that takes up space and has mass.

Great job, Ben! Matter is all around us, and it can be a solid, liquid, or gas. This week we're going to talk about how we can describe matter to others.

Okay! First, though, I do have a question — why would we want to describe matter anyway?

Everything that we can see is made of matter. Knowing how to describe matter is important because it's one way we can communicate what we see, experience, or discover to others. Let's use our imaginations for a minute. Imagine that you're outside playing and you see the most amazing insect you've ever seen! What would you do?

Oh, I'd run inside as fast as I could and tell you or Mom all about it! I'd probably even ask you to come outside quickly so that I could show it to you.

- [x] An apple or orange
- [ ] Knife (adult supervision)
- [ ] Kitchen scale
- [ ] Cutting board

**Weekly materials list**

Right! And remember, all living and non-living things are made of matter, so as you told us about the insect, you'd also be describing the matter that you saw. You might have described its size, color, shape, or behavior. These are all ways to describe what we see. We're going to talk more about describing matter this week — but first, let's play a game!

Name: ______________________________

## Activity directions:

Butterflies are a type of insect. Study the picture of the butterfly. How would you describe this insect?

Now, ask your teacher or a sibling to play a game with you. Tell them that you are going to describe an insect to them and ask them to try to guess what type of insect it is. Describe the butterfly that you saw in the picture — but don't tell them it is a butterfly! Can your teacher or sibling guess that you are talking about a butterfly based on how you describe it to them?

apply it

1. Did your teacher or sibling guess what type of insect you were describing?

2. Was it easier or harder to describe the butterfly than you thought it would be?

3. How did you describe the butterfly?

4. Can you think of any other types of matter that you have described to someone else?

Hello, friend! Did you have fun describing the butterfly in our last lesson? It was a lot harder to describe than I thought it would be, and I realized how important describing matter can be. I'm ready to jump right into learning how to describe matter today. Let's get started, Hannah!

**materials needed**

- An apple or orange
- Knife (adult supervision)
- Kitchen scale
- Cutting board

Alright! When we describe matter, we can share details about it like the color, shape, size, weight or mass, texture, smell, state, and taste. These are all ways to describe the characteristics of matter.

Wait a minute, let's define that word. **Characteristic** (said this way: kār-ĭk-ter-ĭst-ĭk) means a trait or feature of something.

Thanks, Ben! We can also call these characteristics of matter, like size and shape, the properties of matter. **Property** (said this way: prŏp-er-tē) is another word that means a trait or feature something has.

Okay! Well, we know how to describe things like color, shape, and size — so let's talk about some of the other properties we can describe. First, we can record the weight or mass of an object. Remember, weight is the measurement of gravity's force pulling on an object. On the other hand, mass is the measurement of the material in an object. We can also describe the texture of matter — how does it feel or appear? For example, is it smooth, rough, bumpy, or sticky?

Those are all great ways to describe matter — and don't forget we can also use our senses of smell and taste. But remember, never taste or touch something in science if you don't know what it is and you don't have an adult's permission!

Finally, we can record what state the matter is in. Is it a solid, liquid, or gas? Let's practice examining and describing matter today. This will be fun!

## Activity directions:

1. Choose an apple or an orange to examine. Answer questions 1–4 on the worksheet.

2. Use the kitchen scale to measure the mass of the apple or orange in grams or kilograms. Answer question 5 on the worksheet — and don't forget to write the unit of measurement that you used.
3. Ask your teacher to help you cut the apple or orange in half on the cutting board. Observe the inside of the fruit. Is there anything else you observe about this matter? Write your observations on question 6 of the worksheet.
4. With your teacher's permission, smell and taste the fruit. Write down what you observe on questions 7 and 8 of the worksheet.

Name: ______________________________

apply it

1. I chose a ______________________________ to examine.

2. The color was: ______________________________

3. The shape was: ______________________________

4. Describe the object's size:

______________________________

______________________________

______________________________

5. The mass of the object was: ______________________________

6. Once we cut the fruit in half, I observed:

______________________________

______________________________

______________________________

7. It smelled like:

______________________________

______________________________

______________________________

8. It tasted like:

______________________________

______________________________

______________________________

I'm glad we were able to spend some time learning how to describe matter this week. I didn't think it would be important to know how to describe matter at first, but as we've explored, I learned that it is a really helpful skill to have.

I'm glad you had fun, Ben, and I hope you did too, friend! Did you know that being able to describe matter is an important skill for both scientists and explorers?

Actually, I've been learning all about explorers this week. Meriwether Lewis and William Clark were two explorers who were given the huge task of exploring the Louisiana Territory in the year 1804. As Lewis and Clark explored the new territory in America, they created maps of the land and documented the new types of plants and animals that they saw along the way.

In their journals, Lewis and Clark described their travels, sketched new plants and animals they discovered, and described the things they saw along the way. Their journals became a valuable resource for others. In a way, you could say that they created their very own Science Notebooks — just like we're doing!

Lewis and Clark's journals enabled them to share many new types of plants and animals with others who had never seen them before. Their maps would also help other explorers and pioneers settle the new territory in the years to come.

Wow, thanks for telling us about Lewis and Clark, Ben! I'm glad they documented their discoveries to share with others. That gives me an idea! Let's pretend to be explorers ourselves and document something that we find.

Name: ______________________________

apply it

Everything you see, whether living or non-living, is made of matter. Pretend you are an explorer. Go outside and pick an object to describe. Examine the object, then draw it below and describe what you see in at least two sentences.

I think that knowing how to describe matter is a skill that will come in handy as we continue our science adventure together. Matter is all around us because all living and non-living things are made of matter.

This week we examined and described a piece of fruit. Our activity reminded me that the Bible talks a lot about fruit.

That's right, it does! In Galatians 5:22–23 we read,

*But the fruit of the Spirit is love, joy, peace, forbearance, kindness, goodness, faithfulness, gentleness and self-control. Against such things there is no law.*

Forbearance is a big word that means patience. Ben, can you describe what the fruit of the Spirit is?

Sure! A fruit tree produces fruit — and that fruit tells you what kind of tree it is. When we first moved into our house, we noticed a small tree in the yard. During the springtime, the tree had such beautiful flowers! Hannah and I didn't know what kind of tree it was, so Mom told us to watch and wait.

A few months later, the tree had produced apples that we used to make a yummy applesauce! The tree produced fruit, and the fruit told us what kind of tree it was — an apple tree.

In the same way, our lives produce what the Bible calls fruit. Now, this isn't a fruit that you can eat! It is a fruit that displays who you are and Whom you belong to, God.

Right! The fruit of the Spirit is the fruit that our lives produce when we follow Jesus. As we learn and grow in our faith in Him, our lives produce the fruit of love, joy, peace, forbearance, kindness, goodness, faithfulness, gentleness, and self-control. In John 15:8, Jesus said,

*This is to my Father's glory, that you bear much fruit, showing yourselves to be my disciples.*

The fruit of the Spirit can be examined and described in our lives — and its presence shows others that we are Jesus' disciples. A disciple is someone who follows Jesus. Do you notice that a friend or family member is very kind or loving? Be sure to tell them that this is a fruit of the Spirit!

Look up Galatians 5:22–23 in your Bible. If you'd like, you can highlight these verses. Are there any words you do not understand? Ask your teacher to help you look up the definitions in a dictionary, then memorize these verses with your teacher or with a sibling.

Hello, friend! I think Ben forgot what day today is. Shhh, let's stay quiet and see if he remembers that it's time to add a new page to our Science Notebooks!

Wait, what? How could I forget my favorite day of the week? Woohoo — it's time to add a new page to our Science Notebooks! What are we going to add today, Hannah?

Hmm, well, we learned how to describe matter this week. Let's draw a picture of an apple with a magnifying glass over the top of it — like we're examining the apple very closely! I've got a picture of an apple and magnifying glass right here that we can use for an example.

Here's how our Notebooks turned out this week. I made a rainbow magnifying glass for mine!

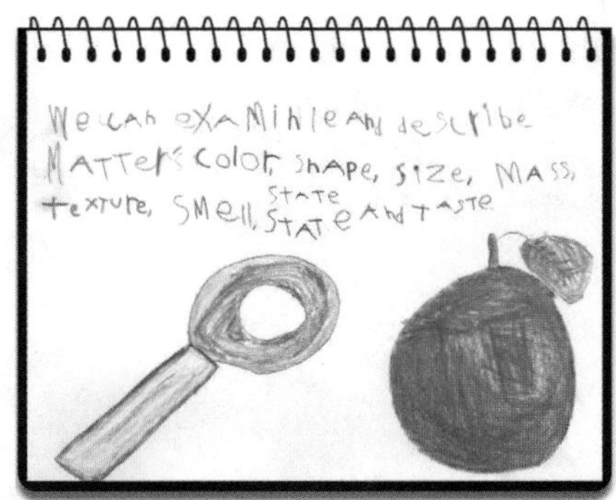

In your Notebook, write: We can examine and describe matter's color, shape, size, mass, texture, smell, state, and taste.

Then draw a picture of an apple and a magnifying glass.

Learning how to examine and describe matter this week reminded us that the fruit of the Spirit can be examined and described in our lives as disciples of Jesus. Copy Galatians 5:22–23 on the back of your Notebook page as a reminder.

*But the fruit of the Spirit is love, joy, peace, forbearance, kindness, goodness, faithfulness, gentleness and self-control. Against such things there is no law* (Galatians 5:22–23).

# Building Blocks of Matter

Day 1

week 8

Hey there, friend, we're excited to be back together for another science adventure! I can't believe how much we've learned together already, and I'm really glad we have so much more to explore.

Me too! We've been learning about matter in our last few adventures. Remember, matter is anything that takes up space and has mass. Everything around us, whether living or non-living, is made from matter. We talked about how to examine and describe matter in our last lesson. But did you know there is actually more to matter than what we can see with our eyes?

Yes, though everything around us is made from matter, matter is made from something much smaller. Hmm, let's think about building blocks for a minute. Have you ever used building blocks to create a tower, friend? I know I have!

Once, I made a very tall tower, and I was so proud of it. From a distance, it looked like my tower was one big piece — but if you looked closer, you would have seen that there were many small building blocks working together to create the tower.

In the same way, when we look at the matter around us, it can appear to be just one object. But if we were able to examine it much more closely, we would discover that there are many smaller parts all working together. We call these smaller parts atoms, and they are the building blocks of matter.

**Weekly materials list**

- [x] 2 bar magnets
- [ ] Chocolate M&M'S®
- [ ] Paper
- [ ] Pencil

**Name:** ____________________

Neat! I have an apple here, and we know the apple is made of matter because it takes up space and has mass. So, could I just use my magnifying glass to examine the atoms in this apple?

That is a really great question, Hannah! No, atoms are far smaller than anything we're able to see with our eyes. But scientists are able to study the behavior of atoms and describe them. We're out of time together for today, but in our next adventure, we'll be taking a closer look at atoms together. See you then!

apply it

1. Can you think of an example of something that has many small parts working together?

2. Have you ever used small parts to build something bigger?

3. If so, what did you build?

**materials needed**

- 2 bar magnets

Oh good, you're here! We were just getting ready to dive into our exploration of atoms — let's get started.

Remember, atoms are extremely small, and we're not able to see them with our eyes alone. For hundreds of years, however, scientists have worked to learn more about atoms through experiments. They've performed many different experiments to learn how atoms behave alone and with other atoms.

These experiments have also helped scientists to develop a model of what atoms look like. A **model** (said this way: mŏd-l) is an example of what something looks like. Let's take a look at a model of an atom today and learn the parts that make up the atom!

An atom is made of three parts: electrons, protons, and neutrons (said this way: noo-trŏns). At the very middle of the atom, we see what is called the nucleus. The **nucleus** (said this way: noo-klē-ŭhs) is a group of protons and neutrons together. The electrons form an orbit around the nucleus, which is the middle of the atom.

Electrons and protons are similar to magnets; they have a charge that can attract or repel something. **Repel** (said this way: rĭ-pĕl) means to push something away.

Let's talk about magnets quickly to help us understand atoms a little better. Magnets have two sides. We call one side the south pole, and the other side is called the north pole. On a magnet, the south pole is usually marked with an "S," and the north pole is marked with an "N." Each pole on a magnet is attracted toward the opposite pole on another magnet, but it will repel the same pole.

So, the magnet's south pole will be attracted to the north pole on another magnet. But the south pole will repel the south pole on another magnet. Hey, let's go get our magnets and give it a try!

Okay, that sounds like fun! Sometimes when we're learning or trying something in science, we create a lab report. Other times, we just write down the observations we made during the activity. Today, let's practice writing down our observations.

## Activity directions:

ASK PARENT FOR HELP

1. Place both bar magnets on a flat surface then try to push the north pole of one magnet toward the south pole of the other magnet. Answer question 1 on the worksheet to record your observation.
2. Next, push the north pole of one magnet toward the north pole of the other magnet. Answer question 2 on the worksheet to record your observation.
3. Answer question 3 on the worksheet. Then push the south pole of one magnet toward the south pole of the other magnet. Answer question 4 on the worksheet to record your observation.

Name: ____________________

apply it

1. What happened when you pushed the north pole of one magnet toward the south pole of the other magnet?

2. What happened when you pushed the north pole of one magnet to the north pole of the other magnet?

3. What would you expect to happen if you pushed the south pole of one magnet to the south pole of the other magnet?

4. Is that what happened?

I hope you had fun working with the magnets in our last lesson, friend! Now, let's get back to talking about atoms. Hannah, what do magnets have to do with atoms?

**materials needed**

- Plain M&M'S®
- Paper
- Pencil

Well, electrons have a negative charge, and protons have a positive charge — it's kind of like the north and south pole on a magnet. Just like we saw with magnets, opposite charges in the atom are attracted to each other. In other words, an electron will be attracted to a proton, but it will be repelled by another electron. These charges are what help the electrons stay in an orbit around the nucleus.

Neat, I'm glad we've been able to learn the parts of an atom together! I'm curious, though, since atoms are so incredibly small, how are we able to see matter? I'm thinking that the atoms must join together to create the structures and objects we can see?

That is exactly right! Usually, atoms don't stay all by themselves. Instead, atoms will link up with other atoms to create a molecule. We'll be talking more about molecules soon. In the meantime, let's create our own model of an atom. Ready for some fun?

## Activity directions:

ASK PARENT FOR HELP

1. Choose three colors of M&M'S® to use for your atom model. You'll need two M&M'S® in each color.
2. Place two of the first color of M&M'S® in the center of the paper then add two of the second color to create the nucleus.
3. Draw an arrow pointing to the first color of M&M'S® then write "protons" to label them. Draw an arrow pointing to the second color of M&M'S® then write "neutrons" to label them.
4. Draw a small circle around the nucleus of the atom and write "nucleus" to label it.
5. Draw two big circles around the nucleus of the atom. Place two of the third color of M&M'S® on the lines as if they were in an orbit around the nucleus.
6. Draw an arrow pointing to one of the M&M'S® on a line and write "electrons" to label it.

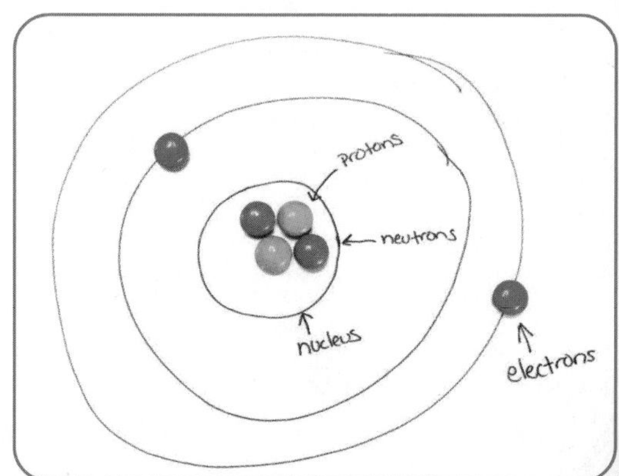

Wasn't it fun to create our own model of an atom? I enjoyed putting it together.

It's always fun to explore a new topic in science together. I think atoms are really interesting, and I enjoy seeing God's wisdom in their design. We'll be talking more about that in the weeks to come. In the meantime, though, our science adventure this week reminded me of Colossians 1:16, which says,

*For in him all things were created: things in heaven and on earth, visible and invisible, whether thrones or powers or rulers or authorities; all things have been created through him and for him.*

I love that verse! The field of chemistry often explores and studies things that we cannot see with our eyes alone. As we explore chemistry, we get to see God's infinite wisdom on display. Infinite is a word that means without limit.

I can see things like the delicate hummingbird, a beautiful flower, or the glowing moon at night. I'm often amazed by the things God created that I can see with my own eyes. But as we explore chemistry and learn about the things God created that we cannot see — like atoms — I'm even more amazed by His power.

God created the things we can see and the things we cannot see — and it's His power that sustains them. Science reminds us that God is wise and powerful. It draws our hearts to praise and worship Him. I'm excited to learn more about chemistry and to see more of God's wisdom on display!

Talk with your family about all of your favorite parts of God's creation. Where do you see God's wisdom and power on display in the world around you? Then look up Colossians 1:16 in your Bible. If you'd like, you can highlight this verse. Memorize Colossians 1:16 with your teacher or with a sibling.

Do you know what day it is?

Oh Ben, that is a silly question! This is your favorite day; you already know what day it is!

True! I just wanted to make sure that you knew what day it is — this is the day we get to add a new page to our Science Notebook! We had so much fun this week learning about atoms, so I thought we could draw a model of an atom in our Notebook this week.

I like that idea!

Sweet! I have an image right here that we can use for an example. Let's get started, and don't forget to show someone your Notebook once you're done and tell them all about what you learned this week.

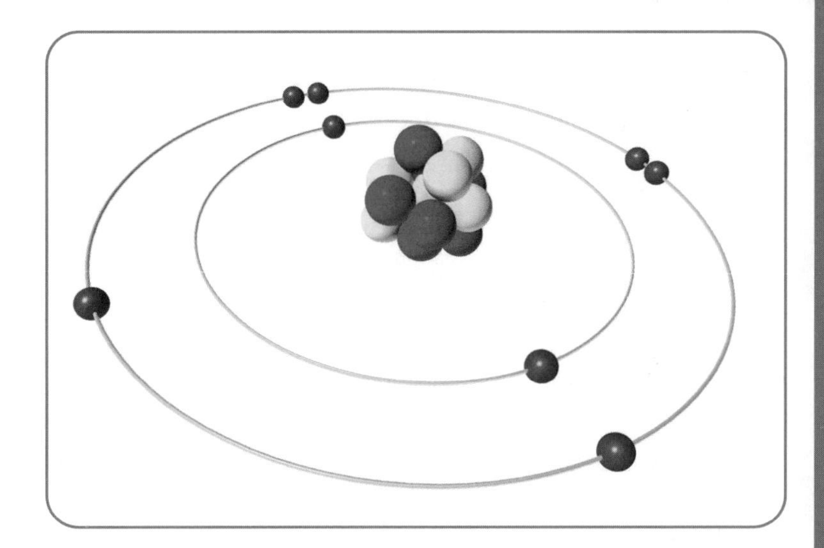

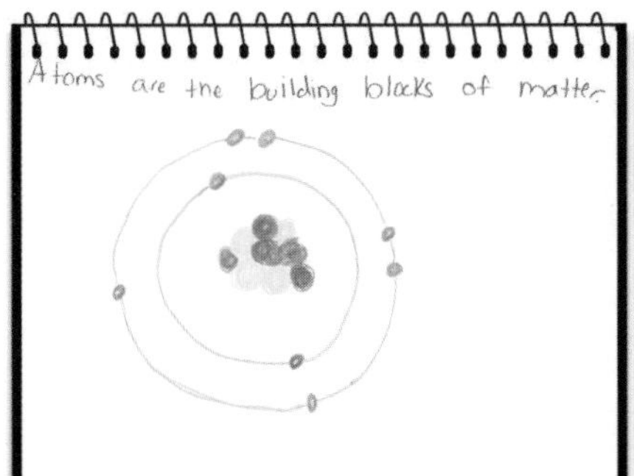


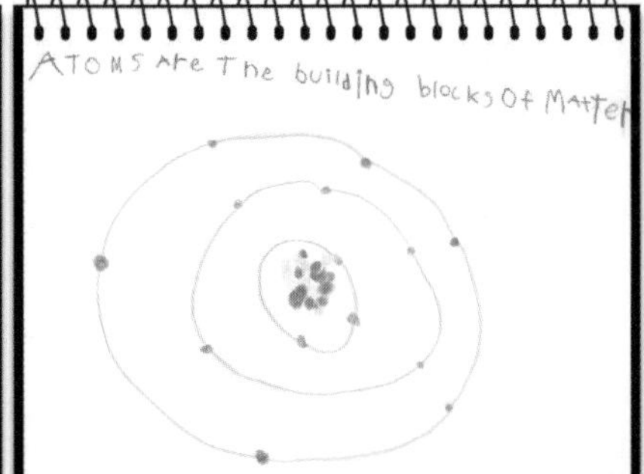


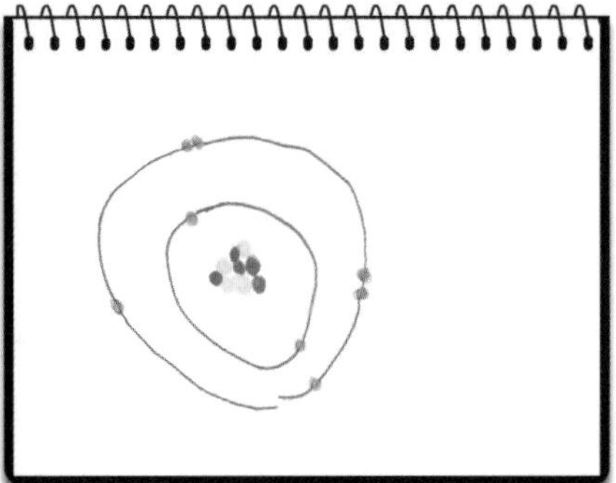

In your Notebook, write: Atoms are the building blocks of matter.

Then draw a picture of an atom.

Learning about atoms this week reminded us that God created and sustains all things. Copy Colossians 1:16 on the back of your Notebook page as a reminder.

*For in him all things were created: things in heaven and on earth, visible and invisible, whether thrones or powers or rulers or authorities; all things have been created through him and for him* (Colossians 1:16).

week 9

# Elements

Day 1

Oh, hey there, friend! I'm glad you're back for another science adventure. Last time, we learned that atoms are the building blocks of matter. Hannah, do you remember the three parts an atom is made of?

I sure do! An atom is made of protons, neutrons, and electrons. Now that we've learned about atoms, it's time to explore the world of elements!

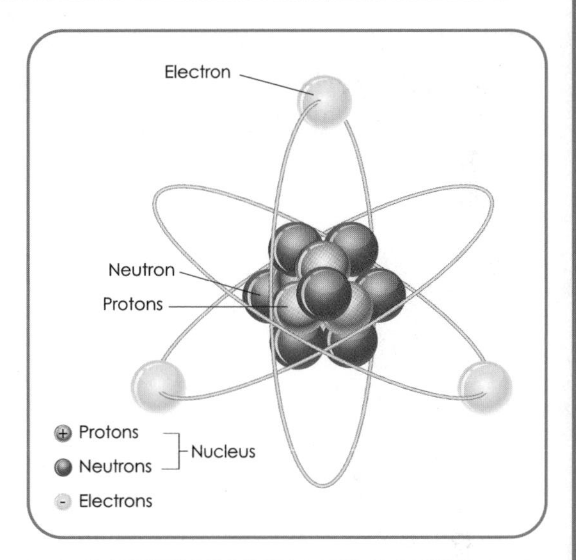


I'm excited to get started! But first, I have a question. Are all atoms the same? In other words, do they all have the same number of protons, neutrons, and electrons?

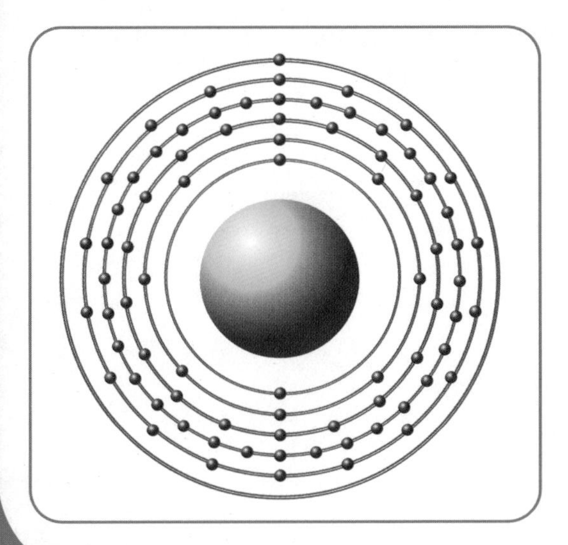

I'm glad you asked. God designed protons, neutrons, and electrons to combine in different ways to create different atoms. Let's talk about a couple different atoms today — we can start with gold.

Gold is a type of metal that is very valuable. Because it is valuable, it is also called a precious metal. Gold is often used to create valuable jewelry, like wedding rings, or it can be used as money. We know that gold is matter because it takes up space and has mass — gold is made of atoms. A gold atom has 118 neutrons and 79 protons in its nucleus (left).

Name: ____________________

Whoa, it would take a lot of M&M'S® to create a model of a gold atom!

It sure would! Now remember, last time we learned that electrons have a negative charge and protons have a positive charge. If an atom has more protons than electrons, the atom would have a positive charge. But if the atom had more electrons than protons, it would have a negative charge. However, an atom prefers to have an equal number of electrons and protons — this gives the atom a neutral charge.

A neutral charge means that the atom is balanced, so it won't have a positive or a negative charge to attract or repel.

Exactly. So, since a gold atom has 79 protons in its nucleus, it also has 79 electrons orbiting around the nucleus. We're out of time for today, but we're going to continue our discussion tomorrow. We'll see you then!

apply it

1. Have you ever seen gold jewelry or gold that is used as money?

____________________

2. Silver is another type of precious metal. A silver atom has 60 neutrons and 47 protons in its nucleus. How many electrons does silver have orbiting the nucleus? Remember, an atom likes to keep protons and electrons balanced.

____________________

3. Silver is used for silverware and also for jewelry. Can you find anything made of silver in your home? Write down what you find.

____________________

____________________

____________________

We're continuing our exploration of atoms this week, and today we're ready to talk about the elements.

Let's get started! Atoms are the building blocks of matter. God designed the number of protons, electrons, and neutrons in an atom to combine in different ways to create different atoms. These different atoms are called substances.

A **substance** (said this way: sŭb-stŭhns) is a certain kind of matter. When a substance is pure and cannot be broken down into any other substances, we call it an **element** (said this way: ĕl-ŭh-mĕnt).

Hmm, that's a little tricky to understand — but I have an example to help us! Imagine we have a peanut butter cracker sandwich in front of us. What is it made from?

That's easy! The sandwich is made from peanut butter spread between two crackers.

Right — the peanut butter and the two crackers combine to create a sandwich. Let's imagine now that we wanted to pull apart the sandwich and separate the ingredients. If we did that, we would have two separate items: the peanut butter and the crackers. We wouldn't be able to separate those items any further.

Ah, in your example, the crackers are like an element, and the peanut butter is like another element.

Right! An element cannot be broken down into any other substances. An element is made from one type of atom. Gold is an element, and pure gold is made from only gold atoms. Silver is another type of element — it is made from only silver atoms.

Scientists have been able to discover 98 elements that are found naturally on the earth. They've also been able to make 20 other elements in a laboratory. This gives us 118 different elements all together. God designed the elements to stay consistent — they are the same no matter where we find them, whether on the earth, deep within the earth, or far into outer space. Let's review what we learned today!

Name: ____________________

Copy each definition below.

1. A substance is a certain kind of matter.

2. An element is a pure substance that cannot be broken down into other substances.

I'm ready to learn more about elements today, how about you?

What are we waiting for? I've got a question to get us started. We learned last time that there are 118 different elements scientists have discovered. How do they keep track of all those different elements?

That's exactly what we are going to talk about today. As scientists began to discover different elements, they knew they also needed to find a way to organize the elements and show the relationships between them. But with so many elements, it was definitely not an easy thing to do!

Many different scientists worked to organize the elements, but one in particular became known as the father of the way we organize them. His name was Dmitri Mendeleev (said this way: Dŭh-mē-trē Mĕn-dŭh-lāy-ŭhv). Mendeleev was born in Russia in 1834, and God gave him a brilliant mind for science.

During Mendeleev's time, scientists only knew of about 63 of the elements. Mendeleev was determined to understand and organize those elements. He asked questions and played with each element like the pieces to a puzzle until the patterns became clearer to him.

But something still wasn't making sense. As he continued to work, Mendeleev realized that there must be more elements that would complete the patterns he was seeing — they just hadn't been discovered yet. He was right. And in fact, God had created the elements with such fine organization that Mendeleev was even able to predict what those missing elements would be like!

Mendeleev's chart organizing the elements became what we now call the periodic table of elements. Eventually, the missing elements were discovered, just as Mendeleev had predicted. Though other scientists also worked to organize the periodic table of elements a little better, Dmitri Mendeleev is known as the father of the periodic table.

The periodic table of elements helps scientists to organize and examine all of the different elements. It's able to give us a lot of information about each element in a clear way. We'll talk a little more about that soon! Mendeleev was able to see God's wisdom, consistency, and organization on display in the elements. Sadly, however, though Mendeleev believed there was a God, he rejected following Christ.

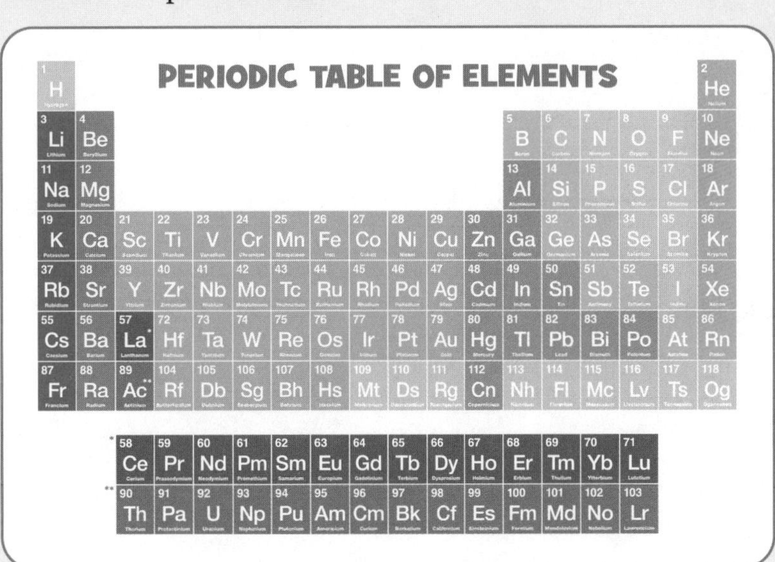

Name: ______________________________

1. There are 118 elements on the periodic table of elements. Find these elements in the word search below.

Gold Oxygen Nitrogen Silver Hydrogen

Sodium Zinc Nickel

| | | | | | | | | | | |
|---|---|---|---|---|---|---|---|---|---|---|
| O | H | Y | D | R | O | G | E | N | X | Z |
| X | Z | N | V | B | S | O | D | I | U | M |
| Y | S | I | L | V | E | R | L | C | F | G |
| G | A | S | N | E | N | E | L | K | B | O |
| E | S | V | Q | C | W | P | O | E | K | L |
| N | I | T | R | O | G | E | N | L | P | D |

2. How does the periodic table of elements help scientists?

# Day

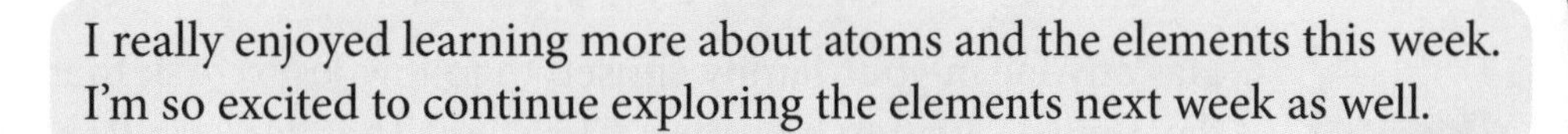


Me too. This week, I've been thinking about how God created each of the elements that scientists have been able to discover. The way the elements can be organized is amazing to me!

At the beginning of our science adventure this year, we talked about how science helps us organize and understand God's creation. It's important to remember that science can be organized because God created the world in an organized way. Science reveals the organized nature, infinite wisdom, and the power of our Creator.

The more we learn together, the more it reminds me of what we read in the Bible in Romans 1:20,

*For since the creation of the world God's invisible qualities—his eternal power and divine nature—have been clearly seen, being understood from what has been made, so that people are without excuse.*

As we study God's creation through science, it reveals His eternal power and divine nature to us. The delicate design of a butterfly's wing, incredible night sky, and even the organization of the elements declare to us the glory, power, and majesty of God.

But like we saw this week, we have a choice to make as we continue to learn. We can choose to see God's glory, power, and majesty on display as we study His creation — or we can reject Him.

Though many scientists do follow Jesus Christ with their lives and trust what the Bible tells us, many others choose to close their eyes to God's glory, power, and majesty in His creation. They refuse to recognize God as our Creator and ultimately choose to reject God in their lives.

I've chosen to follow Jesus for my whole life, and I love to see His amazing design on display when we study science!

Has anyone in your family chosen to follow Jesus? Ask them to tell you about the day they decided to become a Christian. Then look up Romans 1:20 in your Bible. If you'd like, you can highlight this verse. Memorize Romans 1:20 with your teacher or with a sibling.

Hey, friend! We're here and ready to add a new page to our Science Notebook today.

We learned more about atoms, elements, and the periodic table of elements this week. I'm excited to share what we've learned in our Notebooks! What should we draw this week, Hannah?

Well, we talked a little bit about the elements of silver and gold. Mom told us that the silverware in the kitchen is made from silver! I was thinking we could draw a picture of our silverware this week. I have an example picture we can use right here.

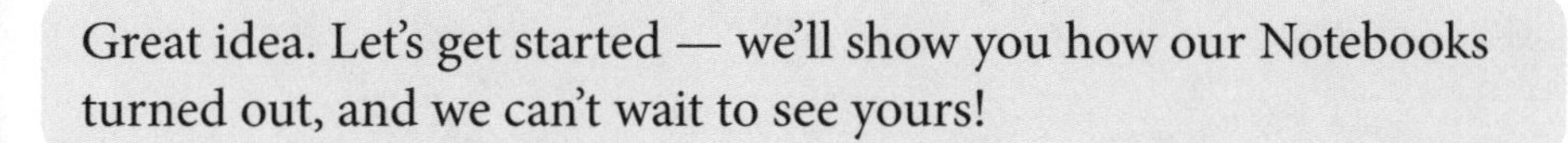

Great idea. Let's get started — we'll show you how our Notebooks turned out, and we can't wait to see yours!

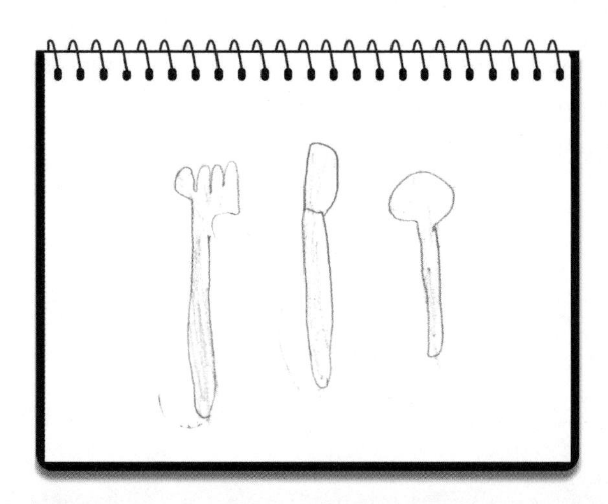

In your Notebook, write: An element is a pure substance that cannot be broken down into other substances. Silver is an element.

Then draw a picture of silverware.

Learning about the organization of the elements this week reminded us that God's power is on display through His creation. Copy Romans 1:20 on the back of your Notebook page as a reminder.

*For since the creation of the world God's invisible qualities—his eternal power and divine nature—have been clearly seen, being understood from what has been made, so that people are without excuse* (Romans 1:20).

# Periodic Table of Elements

Day 1

Hey there, friend! Did you bring some scuba diving gear with you today?

Huh? Why would they need scuba diving gear, Ben?

Because we're diving deeper into the periodic table of elements this week, of course!

Oh, Ben, you're so goofy!

Okay, okay. We're not really going scuba diving today. The phrase "diving deeper" is a figure of speech — it's another way to say that we're going to learn more together. Anyway, let's get started!

In our last adventure, we learned about elements. Remember, an element is a substance that cannot be broken down into any other substances. We also learned about how elements are organized on the periodic table of elements.

Oh, and I forgot to mention before, we can call the periodic table of elements the periodic table for short. Let's take a look at the periodic table again! The first thing I notice is the number in the corner of each square. What does that number mean?

**PERIODIC TABLE OF ELEMENTS**

| | | | | | | | | | | | | | | | | | |
|---|---|---|---|---|---|---|---|---|---|---|---|---|---|---|---|---|---|
| 1 H | | | | | | | | | | | | | | | | | 2 He |
| 3 Li | 4 Be | | | | | | | | | | | 5 B | 6 C | 7 N | 8 O | 9 F | 10 Ne |
| 11 Na | 12 Mg | | | | | | | | | | | 13 Al | 14 Si | 15 P | 16 S | 17 Cl | 18 Ar |
| 19 K | 20 Ca | 21 Sc | 22 Ti | 23 V | 24 Cr | 25 Mn | 26 Fe | 27 Co | 28 Ni | 29 Cu | 30 Zn | 31 Ga | 32 Ge | 33 As | 34 Se | 35 Br | 36 Kr |
| 37 Rb | 38 Sr | 39 Y | 40 Zr | 41 Nb | 42 Mo | 43 Tc | 44 Ru | 45 Rh | 46 Pd | 47 Ag | 48 Cd | 49 In | 50 Sn | 51 Sb | 52 Te | 53 I | 54 Xe |
| 55 Cs | 56 Ba | 57 La* | 72 Hf | 73 Ta | 74 W | 75 Re | 76 Os | 77 Ir | 78 Pt | 79 Au | 80 Hg | 81 Tl | 82 Pb | 83 Bi | 84 Po | 85 At | 86 Rn |
| 87 Fr | 88 Ra | 89 Ac** | 104 Rf | 105 Db | 106 Sg | 107 Bh | 108 Hs | 109 Mt | 110 Ds | 111 Rg | 112 Cn | 113 Nh | 114 Fl | 115 Mc | 116 Lv | 117 Ts | 118 Og |

Great question! That number is called the atomic number. The **atomic number** (said this way: ŭh-tŏm-ĭk nŭhm-ber) tells us how many protons the element has in its nucleus. Let's look at the first element on the top left-hand side of the periodic table.

**Name:** ______________________________

All right, so the element I'm looking at has the number 1 and a big H.

H is the abbreviation of the element's name. Remember, an abbreviation is a shorter way of writing something. On the periodic table, the element's abbreviation is called its symbol. H is the symbol for the element hydrogen (said this way: hī-drŭh-jŭhn). The 1 in the hydrogen square means that an atom of hydrogen has 1 proton in its nucleus.

1
H
Hydrogen

Interesting! I'm excited to learn more together tomorrow. In the meantime, let's look at the periodic table and see how many elements we can recognize from things we've learned about in the past.

apply it

1. Ask your teacher for the periodic table of elements (found in the back of this book). There are 118 elements on the periodic table. You may have learned about some of the elements in the past, and you may recognize ones like silver, gold, oxygen, or nitrogen.

Look at the elements in the periodic table. Are there any elements you've learned or heard about before? If so, write the names of the elements below.

______________________________

______________________________

______________________________

______________________________

______________________________

______________________________

______________________________

______________________________

______________________________

**Note**

There is a periodic table in the back of this book that can be torn out and laminated for easy reference.

2. Pick one of the elements you've learned about before. What is the element's name, symbol, and atomic number? Write them below.

**Name:** ______________________________

**Symbol:** ________ **Atomic number:** ________

Let's examine a few more elements on the periodic table together today.

Okay, I see an element I recognize! Can you find the symbol He on the periodic table? It's on the top of the far right-hand side. The symbol He stands for helium — didn't we use helium (said this way: hē-lē-ŭhm) to inflate balloons for Uncle Gus' birthday party last week?

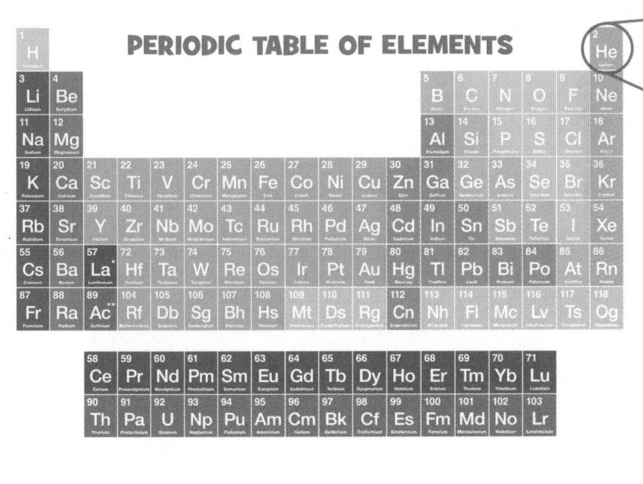


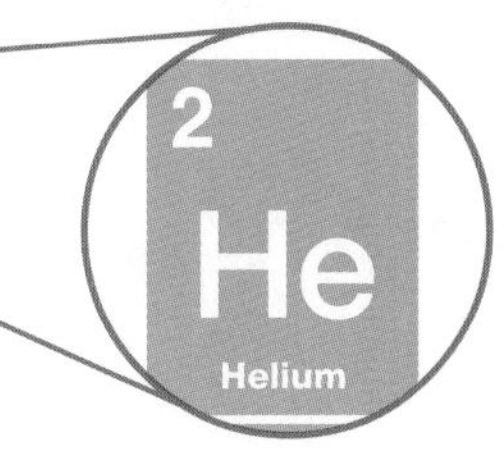


We did! Helium is a type of gas.

I also see the number 2 in the He square. The number two is helium's atomic number, so that means helium has two protons in its nucleus. Hey, Hannah, I also noticed that there are other elements below helium — and they all have the same colored squares. What does that mean?

Great question! Remember, the periodic table helps us to organize the elements and gather information about them quickly. To answer your question, let's learn two new words — vertical and horizontal.

**Vertical** (said this way: ver-tĭ-kŭhl) means that something is in a position of being up and down. For example, when I'm writing with my pencil, my pencil is vertical. **Horizontal** (said this way: hōr-ŭh-zŏn-tl) means that something is in a side-to-side position — like when I lay my pencil back down on the table.

On the periodic table, elements are arranged in horizontal and vertical rows. Elements that are vertical — like helium and the elements underneath it in the same color — are called a group. A group of elements have similar properties to each other. They behave and react to other elements in a similar way.

Helium, neon (said this way: nē-ŏn), argon (said this way: år-gŏn), krypton (said this way: krĭp-tŏn), xenon (said this way: zē-nŏn), and radon (said this way: rāy-dŏn) are a group of elements called noble gases.

Noble, huh? Are these gases like royalty, then?

Good guess, but not quite! Let's say I was trying to make you angry by irritating you. You could choose to react angrily or to simply walk away. If you walked away, we might say that you chose the dignified, or noble, path — you chose not to react. In a similar way, the noble gases don't react with other elements, so that's why we call them noble.

Name: ______________________________

1. The periodic table of elements gives us information on different elements. Can you remember what information it shows us? Write **atomic number**, **symbol**, and **name** on the correct lines to label each part.

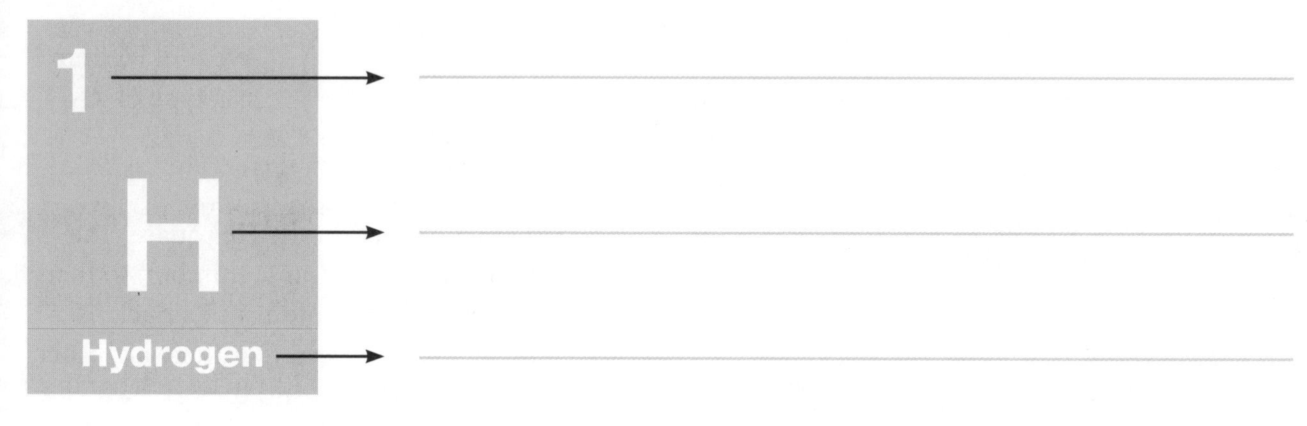


2. Helium, argon, krypton, xenon, and radon are a group of elements called noble gases. Why are these gases called noble?

Hello, and welcome back! We're going to explore the periodic table a little more today. Last time, we talked about a group of elements known as the noble gases. On our example periodic table, these elements are shown in green. Other groups of elements are shown in other colors.

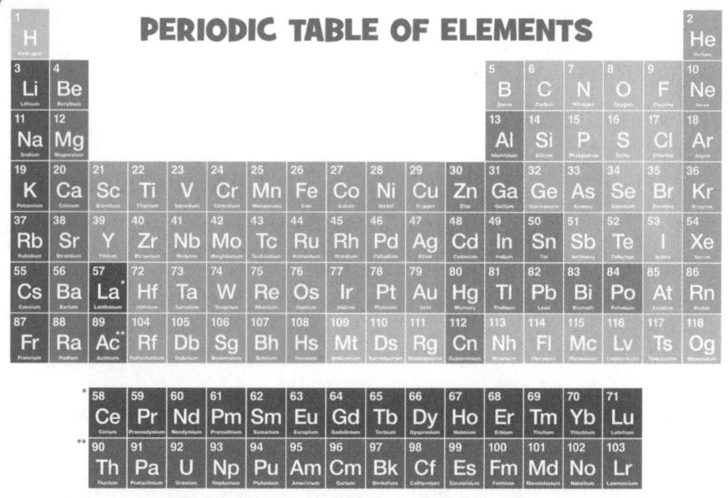


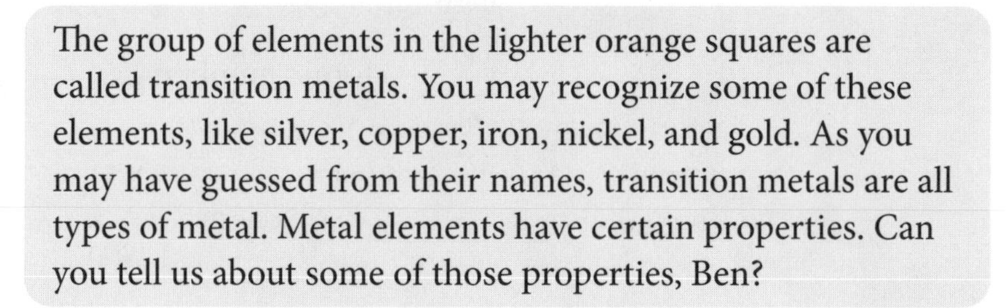


We learned last time that a group of elements have similar properties to each other. They behave and react to other elements in a similar way. Let's talk about another group of elements today. Hmm, let's examine the group of light orange elements. What are those called, Hannah?

The group of elements in the lighter orange squares are called transition metals. You may recognize some of these elements, like silver, copper, iron, nickel, and gold. As you may have guessed from their names, transition metals are all types of metal. Metal elements have certain properties. Can you tell us about some of those properties, Ben?

Sure! Metals usually require a lot of heat to melt them. We can use heat or force to shape metals into useful tools, building materials, jewelry, and more! Metals can also be used to conduct, or transport, heat or electricity. For example, the element of copper is often used in electrical wires.

Thanks, Ben. Copper can also be formed into pipes to transport water through a house. Elements are valuable and can perform many different tasks. Iron is also a transition metal that can be combined with carbon, another element, to create steel.

Oh, and steel is used to create tools, buildings, vehicles, and even some of the appliances we use in our homes.

Iron is also an important element to our bodies. Without iron, our blood system wouldn't be able to carry oxygen throughout our body. The elements God created are useful in so many different ways, and I have a feeling we'll be talking more about them as we continue our science adventure.

We'll learn more about the periodic table in science as we get older. In the meantime, it has been fun to learn more about the elements and how they are organized.

Name: ______________________________

Let's practice finding elements on the periodic table. Use your periodic table of elements (found in the back of this book) to find each symbol below on the periodic table. Then write down the element's name and atomic number.

1. **Cu:** Name of the element: ______________________ Atomic number: ______

2. **Al:** Name of the element: ______________________ Atomic number: ______

3. **O:** Name of the element: ______________________ Atomic number: ______

4. **Fe:** Name of the element: ______________________ Atomic number: ______

5. **Ni:** Name of the element: ______________________ Atomic number: ______

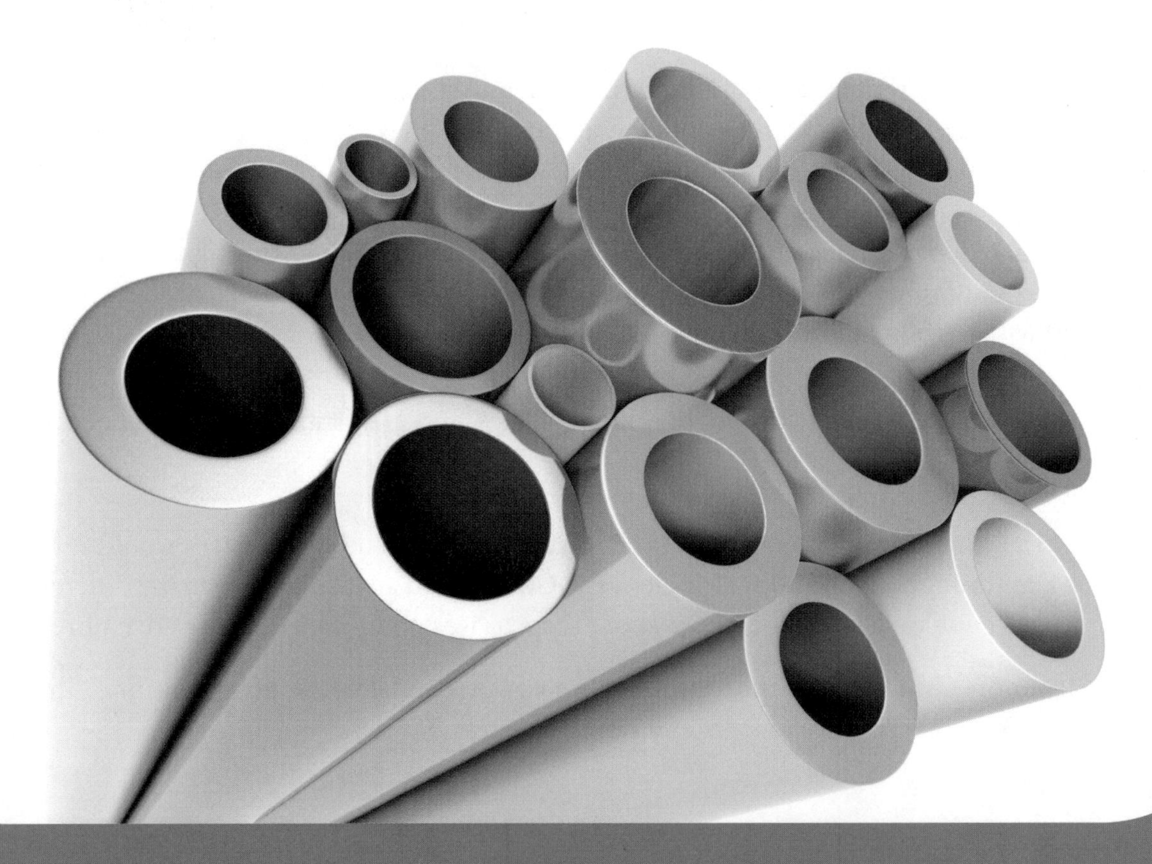

What a fun week it has been as we've learned more about the elements together! I can hardly wait to continue our science adventure with you next week as we begin exploring molecules.

Oh, that sounds like fun! As we've been learning this week, I've been studying the periodic table on my own as well. It's fun to learn the names and properties of different elements. Mom helped me find a book about gold, and it's been interesting to learn more about it. We found the element of gold, on the periodic table last week and learned that it is a transition metal this week.

I remember, the symbol for gold is Au. Gold's atomic number is 79, which means an atom of gold has 79 protons in its nucleus.

Right. Gold was a precious and valuable metal during Bible times, just as it is today. In the book of Genesis, we read about how God created the world and planted the Garden of Eden. In Genesis 2:10–11 it says,

*A river watering the garden flowed from Eden; from there it was separated into four headwaters. The name of the first is the Pishon; it winds through the entire land of Havilah, where there is gold.*

Wow, so there was gold even at the very beginning!

There sure was! God created the elements in His wisdom — they are important and some are even considered precious. It reminded me to think about what I consider to be precious or important in my life. In Sunday school this week, we talked about Psalm 19:9–10, which says,

*The fear of the LORD is pure, enduring forever. The decrees of the LORD are firm, and all of them are righteous. They are more precious than gold, than much pure gold; they are sweeter than honey, than honey from the honeycomb.*

When the Bible talks about the fear of the Lord, it means that we are amazed by God, that we respect Him, and that we follow His ways. God's ways are firm and righteous — we read and learn about them in the Bible. The directions God gives us in the Bible help us to live our lives the way He designed. They protect us. The Bible is God's Word to us, and it is more precious than gold.

I'm so glad you shared those verses, Ben! Let's memorize them together.

Look up Psalm 19:9–10 in your Bible. If you'd like, you can highlight these verses in your Bible. Memorize Psalm 19:9–10 with your teacher or with a sibling.

Grab your Science Notebook, it's time to add a new page!

Woohoo! I have my art supplies right here. We explored a few of the elements on the periodic table this week. I was thinking we could pick one element and add it to our Notebook.

Ah, I like your idea! We could draw a square then write the atomic number, symbol, and name of the element. Then we could color the background the same color that the periodic table of elements shows for that element.

Let's get started! Look at your copy of the periodic table of elements and pick out one of the elements we talked about this week, like copper, neon, hydrogen, or iron. Or you can choose an element you'd like to learn more about. Here is how each of our notebooks turned out. Hannah picked the element of iron. I drew helium, and Sam picked copper. Have fun creating your picture!

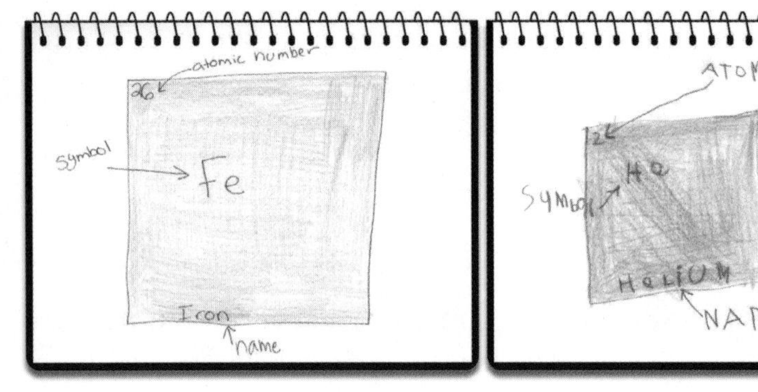

In your Notebook, draw a picture showing one of the elements from the periodic table.

Then draw an arrow pointing to the element's symbol and label the arrow "symbol." Draw an arrow pointing to the atomic number and label it "atomic number." Draw an arrow to the element's name and label it "name."

Learning about valuable and precious elements this week also reminded us that the Bible is more precious than gold. Copy Psalm 19:9 on the back of your Notebook page as a reminder.

*The fear of the LORD is pure, enduring forever. The decrees of the LORD are firm, and all of them are righteous* (Psalm 19:9).

week 11

# Molecules

Welcome back! We're excited to begin another science adventure with you. I've had so much fun learning about atoms and the elements together. But I've been wondering, do atoms always stay by themselves?

Good question. Let's explore the answer together this week! It is actually very rare for atoms to stay all by themselves. Instead, two or more different atoms can combine to create many different substances. When two or more atoms combine, or bond together, we call them a **molecule** (said this way: mŏl-ŭh-kyool).

Now I have another question. When we put two small building blocks together, it creates a bigger block. I would imagine it is the same with atoms? When two small atoms combine, they would create a bigger structure — the molecule. Since the molecule would be bigger than one single atom, are we able to see molecules?

We can't see molecules with our eyes alone, but scientists can use very powerful microscopes to examine them. Though you and I can't see the molecules themselves, we can see the substances that they create. We're going to examine some of those substances this week — and even create our own molecule model!

Let's start with one very important gas: oxygen. Oxygen is an element, and it is part of the air we breathe in. Oxygen is essential for our bodies to function and remain alive.

On the periodic table of elements, oxygen's symbol is O, and its atomic number is 8.

- Marshmallows
- Toothpicks
- Permanent marker
- 2 Tbsp dirt
- 2 bowls
- Hydrogen peroxide

**Weekly materials list**

When two oxygen atoms bond, they form an oxygen molecule. If only two atoms bond together, we call the molecule a **diatomic** (said this way: dī-ŭh-tŏm-ĭk) molecule.

Ah, so an oxygen molecule is a diatomic molecule. I have an easy way for us to remember that word. The word "di" means two. When "di" is placed at the beginning of a word, it tells us that there are two of something. So, "di" means two, and "atom" or "atomic" tells us that there are two atoms.

Name: ______________________

## Activity directions:

**materials needed**

- ☐ Marshmallows
- ☐ Toothpicks
- ☐ Permanent marker

1. An oxygen molecule is made from two oxygen atoms. Let's create a model of an oxygen molecule together. We're going to pretend the marshmallows are oxygen atoms.
2. Pick out two marshmallows. Then carefully use the permanent marker to write "O" on each marshmallow. Remember, O is the symbol for an oxygen atom.

3. Poke a toothpick into one marshmallow then poke the other end of the toothpick into the second marshmallow.
4. The toothpick joins, or bonds, the marshmallow oxygen atoms together to create a model of a diatomic oxygen molecule.

Fill in the blanks with the missing words: **molecule, diatomic**

1. If only two atoms bond together, we call the molecule a ______________________ molecule.
2. When two or more atoms combine, or bond together, we call them a ______________________.

We're back together, and I have another question to get us started today. Yesterday, we learned that atoms usually don't exist all by themselves; they bond together with other atoms. We also learned that when two atoms bond together, we call the result a diatomic molecule. Hannah, can more than two atoms or two different atoms also bond together?

**materials needed**

- Dirt
- Tablespoon
- 2 bowls
- Water
- Hydrogen peroxide

Absolutely! Diatomic molecules are formed by two of the same type of atoms bonding together — like two oxygen atoms. But when two or more different atoms bond together to form a molecule, we call it a **compound** (said this way: kŏm-pound). Hmm, let me think of an example. . . . Ben, what did you sprinkle on your cereal this morning?

Sugar! Is sugar a compound?

It sure is! A sugar molecule is made from 11 oxygen atoms, 12 carbon atoms, and 22 hydrogen atoms all bonded together.

Hmm, I've got another compound for you. This compound is something we drink every single day — in fact, our bodies need it to survive.

Let me guess — water!

That's it! A water molecule is formed by the bond of two hydrogen atoms and one oxygen atom. Water is essential to life on earth, and it wouldn't be possible without the bond of those two important elements.

Eek, ouch! I was writing down notes on this piece of paper, and I just gave myself a paper cut!

Ow! That looks deep. We should ask Mom to get you the hydrogen peroxide to clean out the cut. We'll be right back! In the meantime, you can get started on our activity for today.

## Activity directions:

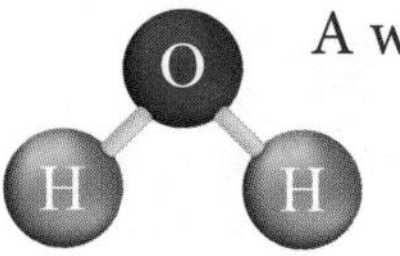


A water molecule is made from two hydrogen atoms and one oxygen atom bonded together, while a molecule of hydrogen peroxide is made from two hydrogen atoms and two oxygen atoms.

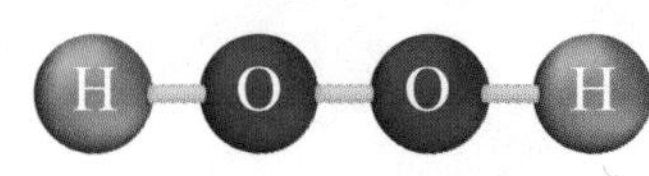

Hydrogen peroxide is a compound that reacts with many things. Hydrogen peroxide's reaction is often used to help kill any germs in a wound. Though water and hydrogen peroxide are both made of hydrogen and oxygen atoms, hydrogen peroxide behaves very differently from water because it has two oxygen atoms instead of one. Let's observe the behavior of water and hydrogen peroxide.

1. Write your name and the date on your lab report on the next page. Next, write the question: How do water and hydrogen peroxide behave differently?
2. Place 1 tablespoon of dirt into each bowl. We're going to add water to one bowl and hydrogen peroxide to the other bowl. What do you think the hydrogen peroxide will do? What do you think the water will do? Write your answers in the Hypothesis section of your lab report.
3. Pour 1 tablespoon of water into the first bowl. How does the water react with the dirt? Write your observation in the "Things I observed" section of your report.
4. Pour 1 tablespoon of hydrogen peroxide into the second bowl. How does the hydrogen peroxide react with the dirt? Write your observation in the "Things I observed" section of your report.
5. Once you've finished observing the reaction of water and hydrogen peroxide with dirt, finish your lab report and record the results of your experiment.

When hydrogen peroxide reacts with something else, the bonds holding the hydrogen and oxygen atoms together begin to break down. This causes bubbles of oxygen to be released.

Name ______________________ Date ______________

# Lab Report

**Question**

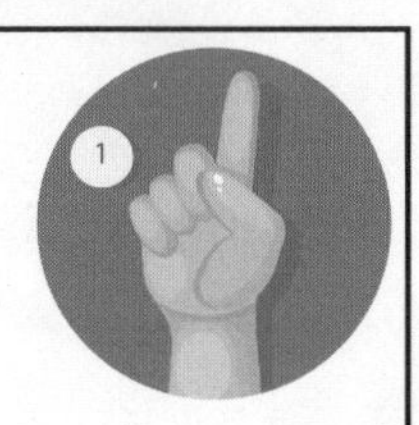


**Hypothesis**

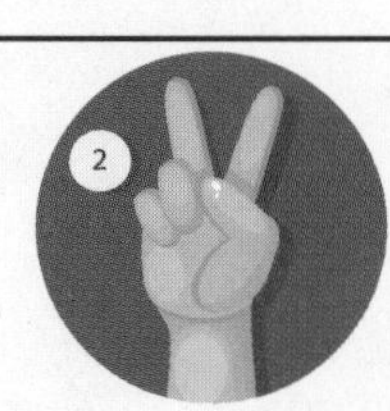


**Things I observed:**

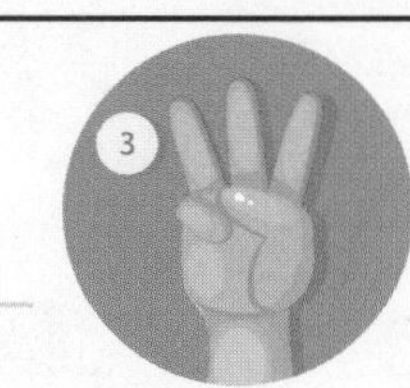


## Results

**What happened in the experiment?**

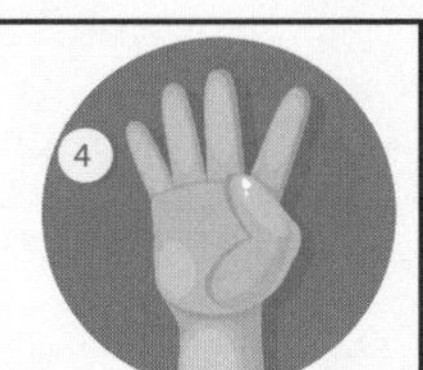


**Was my hypothesis correct?**

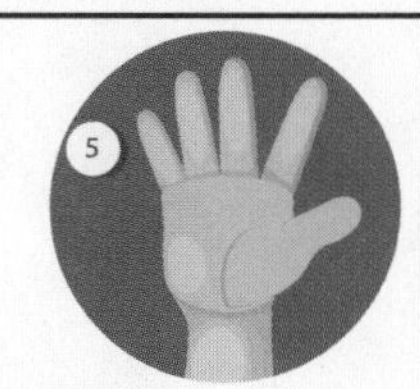

# Additional Lab Notes

Hey there! My paper cut is healing up quickly — I'm glad we have hydrogen peroxide to help make sure a cut like that doesn't become infected and even more painful!

For sure. It is amazing to me how atoms can combine in different ways to create so many different compounds. The compounds they create can even be very different than the elements that bond together to create them! What an amazing system God designed.

We've learned before that organization is very important in science. Being able to share information is also important. Hannah, with so many ways for atoms to combine and make thousands of different compounds, how do you think scientists are able to tell each other what atoms make a compound?

Well, creating models for every compound and taking a picture would take a lot of time. It would also take a lot of time to write out every combination — like sugar, for example. A molecule of sugar is made from 45 atoms bonded together! Scientists could say something like, "A molecule of sugar is 11 oxygen atoms, 12 carbon atoms, and 22 hydrogen atoms." But that doesn't seem like a very organized or easy way to share information about a compound with someone else. It uses a lot of words!

Good observations! That is why scientists use chemical formulas to share what elements are bonded together to form a molecule. Let's look at a molecule of oxygen, for example. Oxygen is made from two oxygen atoms bonded together. Remember, the symbol for oxygen on the periodic table is O.

The chemical formula for oxygen is $O_2$. The O tells us that this molecule is made from oxygen atoms, and the 2 tells us that there are two atoms of oxygen.

Interesting! Let's look at water next. What would be the chemical formula for a molecule of water?

Name: ______________________________

Water is made from two hydrogen atoms and one oxygen atom. The symbol for hydrogen is H, and the symbol for oxygen is O. To write the chemical formula for water, we would write $H_2O$.

I see! The H stands for hydrogen, and the 2 tells us there are two hydrogen atoms. The O stands for oxygen, and since there is just one atom of oxygen, there isn't another number. Let's practice writing some chemical formulas!

O H H

apply it

1. A molecule of sugar is 12 carbon atoms, 22 hydrogen atoms, and 11 oxygen atoms. The chemical formula is $C_{12}H_{22}O_{11}$. Copy the formula below.

2. What do the numbers stand for in a chemical formula?

3. Circle the correct answer. In a chemical formula, the letters tell us:

   a. How many atoms there are.

   b. The symbol of the elements that are bonded together.

4. Bonus! Hydrogen peroxide is made from two hydrogen atoms and two oxygen atoms. How would you write the chemical formula?

I've had a blast learning about molecules and chemical formulas with you this week! I'm also looking forward to building on what we've learned as we explore more chemistry together in the weeks to come. Speaking of building, I've been thinking about how atoms bond together to build many different compounds. Compounds, like water, are so very important to life on earth!

The way atoms join and build together reminds me of the verses in Colossians that Dad read to us at dinner the other night. Let's read Colossians 2:6–7 together:

*So then, just as you received Christ Jesus as Lord, continue to live your lives in him, rooted and built up in him, strengthened in the faith as you were taught, and overflowing with thankfulness.*

Ah, I remember! We have received Christ, and we follow Him — but that isn't the end of our story. Colossians 2:6–7 reminds us that we must keep learning, growing, and building our faith. We must keep our roots strong in Christ. As we learn and grow, it also builds our faith stronger.

But that still isn't all! In Ephesians 4:15–16, the Bible compares those who follow Jesus to a body,

*Instead, speaking the truth in love, we will grow to become in every respect the mature body of him who is the head, that is, Christ. From him the whole body, joined and held together by every supporting ligament, grows and builds itself up in love, as each part does its work.*

We all work together as we follow Jesus, and He's given us each a job to do in the body of believers — just as the hands, feet, eyes, and ears have jobs in our own bodies. As we learn and grow in our faith, we can also help others build their own faith by teaching them what we've learned.

Just like Mom and Dad teach us! Hey, that gives me an idea! You and I can help teach Sam what we've learned from the Bible. Who do you think you can help teach, friend?

Read and discuss 1 Corinthians 12:12–21 with your family. Then look up Colossians 2:6–7 in your Bible. If you'd like, you can highlight these verses. Memorize Colossians 2:6–7 with your teacher or with a sibling.

Do you have your art supplies ready, friend? It's time to add a new page to our Science Notebook! What are we going to add this week, Ben?

Hmm, we explored molecules this week, so I was thinking we could draw a model of a molecule. Let's draw a picture of water and a model of a water molecule!

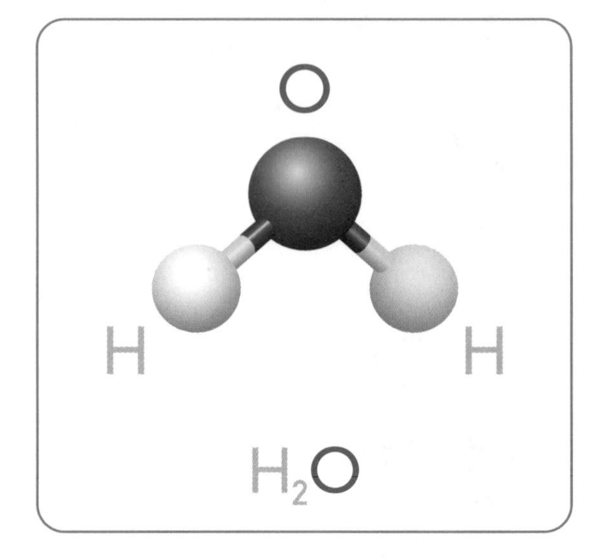


Ah, great idea! We could draw a picture of a glass of water, or a lake — you can use your creativity. Then we'll want to draw a model of a water molecule. I have an example of a water molecule right here.

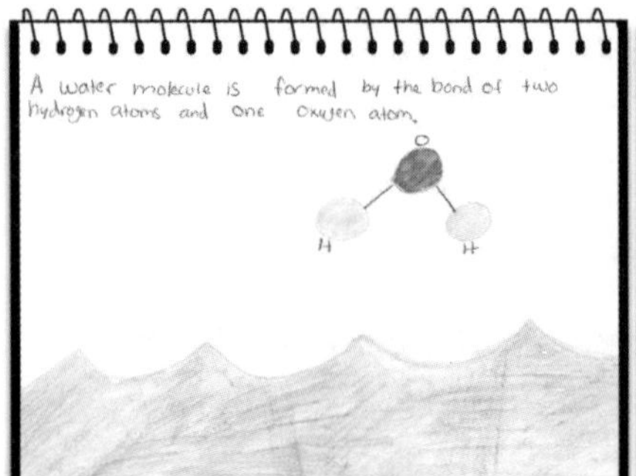

In your Notebook, write: A water molecule is formed by the bond of two hydrogen atoms and one oxygen atom.

Then draw a picture of water and a water molecule model.

Learning about atoms bonding together to build compounds this week also reminded us that we can build up our faith — and help others build theirs. Copy Colossians 2:6–7 on the back of your Notebook page as a reminder.

*So then, just as you received Christ Jesus as Lord, continue to live your lives in him, rooted and built up in him, strengthened in the faith as you were taught, and overflowing with thankfulness* (Colossians 2:6–7).

week 12

# The Carbon Cycle

Day 1

Why, hello there! We've been exploring atoms, elements, and molecules together lately. This week, we're going to turn our focus a little bit and explore some of the cycles God put in place to keep His creation in balance. Before we begin, though, we need to make sure that we understand the meaning of the word cycle.

A **cycle** (said this way: sī-kŭhl) is a process or series of events that happens over and over again. For example, each morning I wake up, get dressed, eat breakfast, and brush my teeth — this is a cycle I follow. Another example of a cycle would be how the sun rises each morning and sets at night in a predictable pattern.

God created cycles for His creation. Some cycles, like the sun rising and setting, give us time and seasons we can follow. Other cycles work to keep creation balanced. Can you think of any other examples of a cycle in creation, Hannah?

Hmm, what about the water cycle? The water cycle keeps water from being used up from the earth.

We've learned about the water cycle in the past — let's review it together quickly! God designed the water cycle to allow water to be recycled through evaporation, condensation, and precipitation. First, the heat from the sun evaporates water from the surface of the earth and from the leaves of plants. When water evaporates from plants, it's called transpiration.

And don't forget water also evaporates from you when you exhale and when you sweat. When water evaporates, it becomes water vapor that travels through the air as very tiny droplets of water.

Next, as water vapor travels higher into the atmosphere, it begins to cool and condense, which forms clouds full of water droplets or ice crystals. Eventually, the water droplets become too big and heavy. When this happens, the water droplets fall out of the cloud as raindrops, hail, sleet, or snow — we call this precipitation!

- Marshmallows ✓
- Toothpicks
- Permanent marker

**Weekly materials list**

Name: ______________________________

As the water returns to the earth through precipitation, some raindrops will fall into streams, rivers, lakes, and the ocean, while other raindrops will fall on land. We call this collection. Some of the water collected will evaporate again, and the process of evaporation, condensation, and precipitation starts all over.

Wait, Mom is calling us for lunch now, so we'll have to continue our discussion tomorrow. See you then!

**apply it**

Fill in each blank with the correct word: **cycle, transpiration, evaporates**

1. When water evaporates from plants, it's called ______________________________ .

2. A process or series of events that happens over and over again is called a ____________ .

3. The sun ______________________ water from the surface of the earth.

4. What do you think would happen to water on the earth if God had not created the water cycle?

______________________________________________________________

______________________________________________________________

______________________________________________________________

______________________________________________________________

______________________________________________________________

**materials needed**

- Marshmallows
- Toothpicks
- Permanent marker

Yesterday, we talked about the water cycle. Over and over again throughout all of history, water has repeated this cycle of evaporation, condensation, and precipitation. I think it's really amazing how God designed this system to recycle water on the earth!

But that isn't the only cycle God put in place on the earth. We're going to explore the carbon cycle together next — but first, we ought to learn a little about the element of carbon.

We've mentioned carbon in a few of our adventures. Carbon can be combined with iron to create steel. Oh, carbon is also part of a sugar molecule. Mom told me before that diamonds are also made from carbon atoms — but sugar and diamonds aren't alike at all!

That is one of the amazing things about the way God designed the elements to combine and bond. The elements on the periodic table can bond together in different ways to create thousands of diverse and unique compounds!

The symbol for carbon is C. Can you find carbon on the periodic table?

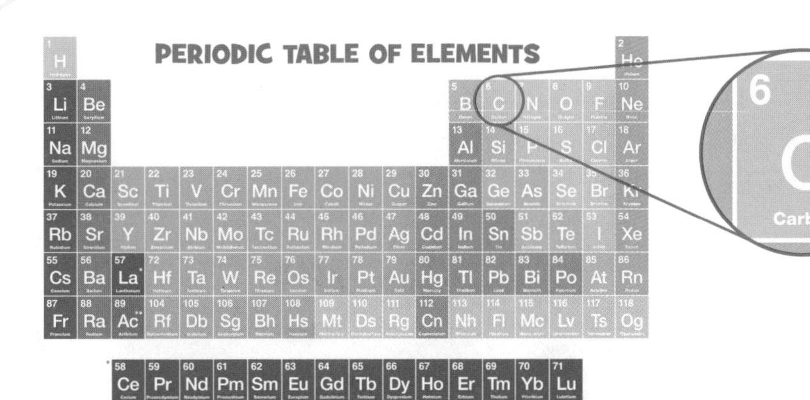


Carbon's atomic number is 6. That means carbon has 6 protons in its nucleus. It looks like carbon is part of the yellow group of elements. The key at the bottom of the periodic table tells us that yellow is for reactive nonmetal elements. The word nonmetal tells us that carbon isn't a metal, like silver or iron. Carbon also reacts with some other elements.

Carbon is an element that is found in all living things and in many nonliving things. Can you give us a few examples of where we would find carbon, Hannah?

Absolutely! Let's start with your pencil. The tip of your pencil is made of a substance called graphite. When carbon atoms bond with each other in layered rows, they form graphite. Graphite is made from just carbon atoms. Diamonds are also made from pure carbon atoms. And don't forget we've already talked about steel and sugar.

You may also recognize carbon dioxide gas. We exhale carbon dioxide, and plants absorb it for photosynthesis (said this way: fō-tō-sĭn-the-sĭs) — we'll talk more about that next time. In the meantime, carbon dioxide molecules are created by one carbon atom bonded to two oxygen atoms, and the chemical formula is $CO_2$. Let's create a model of a carbon dioxide molecule!

## Activity directions:

ASK PARENT FOR HELP

1. Pick out three marshmallows. Then carefully use the permanent marker to write "C" on one marshmallow and "O" on two marshmallows. Remember, C is the symbol for carbon and O is the symbol for oxygen.
2. Place the "C" marshmallow in the middle then poke a toothpick into one side. Attach an "O" marshmallow to the other side.
3. Poke a toothpick into the other side of the "C" marshmallow then attach the other "O" marshmallow. Be sure to show someone your model and explain the carbon dioxide molecule to them!

Hey there! We're continuing to explore the element of carbon this week. Carbon is an important element found in all living things and many nonliving things. In fact, carbon helps to form millions of different compounds!

And just like the water cycle, God designed a cycle for carbon so that it isn't used up from the earth. We call it the carbon cycle — let's learn about it!

We created our own model of a carbon dioxide molecule in our last adventure — and carbon dioxide gas is exactly where we'll begin today. Carbon dioxide is the gas that we exhale, and it's where the carbon cycle begins. You see, plants absorb carbon dioxide gas and use it in the process of photosynthesis.

You may remember from past science adventures that photosynthesis is the process plants use to convert sunlight and carbon dioxide gas from the air into sugar that the plant uses and oxygen that people and animals can inhale.

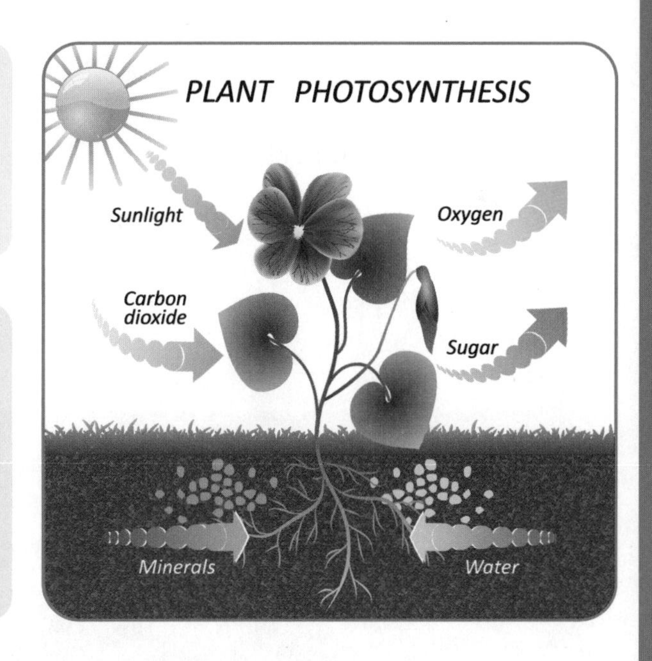


When a person or animal eats a plant for food, they'll also receive some carbon from the plant. This carbon may be digested and returned to the soil as waste or exhaled to be used again.

All living things, like plants, animals, insects, fish, and people, have carbon. When a living thing dies, it will begin to decay.

Hannah and I learned about decay in *Adventures on Planet Earth* — let's review what it means. **Decay** (said this way: dē-kā) means that something is breaking down. We also call this **decomposition** (said this way: dē-cŏm-pō-zĭ-shŭn). Decomposition is the process in which dead material breaks back down into the soil.

Thanks, Ben! During decomposition, carbon is returned to the air or the soil to be used again. For example, think about the leaves on our trees. During the summer, the leaves absorb carbon dioxide for photosynthesis. But in the fall, the leaves die and fall from the tree. Then during the fall and winter, the leaves decompose and return carbon to the air and soil. That carbon will be used by new leaves and plants in the springtime!

Name: ____________________

Follow each prompt and write what you find on the lines below.

1. Find something or someone that exhales carbon dioxide.

____________________

2. Find something that absorbs carbon dioxide.

____________________

3. Find something or someone that consumed carbon from plants today.

____________________

4. Bonus! Can you find old plant material, like leaves or vegetables, decomposing outside?

____________________

I sure enjoyed learning about the carbon cycle with you this week. Carbon is definitely an amazing element. God created it to be so important to life, created a cycle to preserve it, and made it versatile to form millions of different compounds!

Chemistry is definitely an interesting — and fun — field of science. But you know, as we were talking about the carbon cycle this week, it also reminded me that the world was broken through sin. We learn in Genesis chapters 1 and 2 that God designed creation perfectly at the beginning. God created Adam and Eve as the first two people on the earth, and He gave them a command:

*And the LORD God commanded the man, "You are free to eat from any tree in the garden; but you must not eat from the tree of the knowledge of good and evil, for when you eat from it you will certainly die"* (Genesis 2:16–17).

But then in Genesis chapter 3, we read that the serpent deceived Eve, and she ate the fruit against God's directions. Adam also ate the fruit with her. Their choice to disobey God's command brought sin into creation. Sin separates us from God. It always breaks and destroys. Sin corrupted and broke God's original perfect creation. Death and decay entered creation as part of the consequences of sin.

But thankfully, that isn't the end of the story! The Bible reveals God's plan to redeem us from our sin through Jesus. Redeem is a word that means to pay the price of. Because Jesus paid the penalty of sin for us, which is death, we can have a relationship with God again. When we believe and trust in Jesus, He also promises that we will live with Him forever — for all eternity — one day.

In the Book of Revelation, the Bible tells us that one day there will be a new heaven and a new earth where things will be perfect once more. In Revelation 21:3–4 it says,

*And I heard a loud voice from the throne saying, "Look! God's dwelling place is now among the people, and he will dwell with them. They will be his people, and God himself will be with them and be their God. 'He will wipe every tear from their eyes. There will be no more death' or mourning or crying or pain, for the old order of things has passed away."*

As we study science, we do see things like death and decay that remind us of the consequences of sin. But they also remind us that through Jesus, we don't have to suffer the worst consequence of sin: separation from God. Through Jesus, we can have a relationship with God and live with the hope that one day things will be restored.

Do you have a relationship with God? If you do, that is wonderful! If you would like to have a relationship with God, be sure to talk to your parent — they can help you begin your relationship with God today.

Look up Revelation 21:3–4 in your Bible. If you'd like, you can highlight these verses in your Bible. Memorize Revelation 21:3–4 with your teacher or with a sibling.

Well, it's time to add a new page to our Science Notebook today! I've been waiting for you and Hannah so that we can get started.

I'm sorry to keep you waiting, Ben. Let's get right to work. This week, we learned about the carbon cycle and how important carbon is to life on earth. I was thinking we could draw a picture showing some living things like plants and animals since they all would have carbon.

I like it! I have a picture here that we can use as an example. This picture has a bear walking near a pond, but you could add any person, animal, fish, bird, or insect you'd like to your drawing as well.

Here is how our Science Notebooks look. Ben decided to draw the ocean with some fish. Sam and I drew the bear walking near a pond — we even added some colorful fish. We can't wait to see how your Notebook turns out. Use your creativity and have fun!

In your Notebook, write: Carbon is an element that is found in all living things and in many nonliving things.

Then draw a picture of living things.

Learning about the carbon cycle this week also reminded us that creation was broken by sin. Copy Revelation 21:4 on the back of your Notebook page as a reminder that someday God will restore things.

" *'He will wipe every tear from their eyes. There will be no more death' or mourning or crying or pain, for the old order of things has passed away"* (Revelation 21:4).

week 13

# Mixtures & Solutions

Day 1

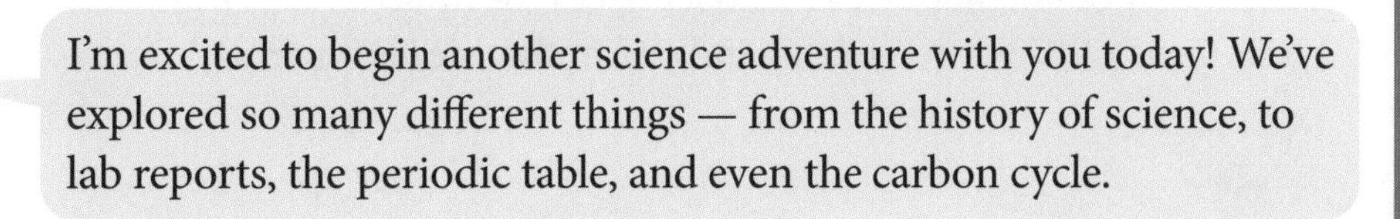

I'm excited to begin another science adventure with you today! We've explored so many different things — from the history of science, to lab reports, the periodic table, and even the carbon cycle.

We've been learning a lot about chemistry! Do you remember when we talked about the elements, Ben?

Yep! Elements are pure substances that cannot be broken down into any other substances. We also talked about molecules and compounds. Remember, compounds are two or more different atoms bonded together to form a molecule. For example, two hydrogen atoms and one oxygen atom bond to form $H_2O$, or water. But compounds aren't the only way things can bond together!

This week, we're going to explore mixtures and solutions. Let's start with mixtures! When two or more substances join, or mix together, we call it a **mixture** (said this way: mĭx-cher).

In the field of chemistry, mixtures have certain characteristics. First, a mixture is made from two or more substances. Second, even after these substances are mixed together, they can still be separated from each other. For example, we can separate mixtures through a fine filter or by boiling.

Hmm, it makes me think of chocolate milk. To create a mixture of chocolate milk, we would mix together milk and sweetened cocoa powder — but if we had a good filter, we could filter the cocoa back out.

Great example! Another characteristic of a mixture is that it doesn't create a new substance. Let's think about how atoms bond to form compounds again. Hydrogen and oxygen bond to create a brand new substance — water! But in Ben's example, chocolate milk isn't a totally new substance — it's simply the combination of milk and sweetened cocoa powder.

- [x] 2 clear glasses or coffee mugs of hot water
- [ ] 2 tea bags
- [ ] Sugar & milk (optional)
- [ ] 1 Tbsp olive or coconut oil
- [ ] Sticky tabs
- [ ] Spoon
- [ ] 2 cups and water
- [ ] 2 tsp salt
- [ ] Stopwatch

**Weekly materials list**

Name: ______________________________

Finally, we learned before that atoms must combine in certain ways in order to create a specific molecule. When two hydrogen atoms and one oxygen atom combine, it creates a water molecule. But if one more oxygen atom combines, it will create hydrogen peroxide, which is very different from plain water. In a mixture, however, things can mix together in different combinations and measurements.

Like chocolate milk — I can mix a lot of sweetened cocoa powder into a small cup of milk or mix a little bit of cocoa powder into a big cup of milk! Either way, I can make chocolate milk. It will just taste a little different depending on the measurements I use.

1. What is a mixture?

______________________________

______________________________

______________________________

______________________________

______________________________

2. Have you ever created a mixture in the kitchen, like chocolate milk? If so, what mixture did you make?

______________________________

______________________________

Circle the word that makes the sentence true.

3. A mixture **does / does not** create a totally new substance.

4. Substances **can / cannot** combine in different measurements to form a mixture.

5. A mixture **can / cannot** be separated once mixed.

Name: ______________________

We're having fun talking about mixtures this week. Do you remember what a mixture is? A mixture is two or more substances mixed together.

We also learned last time that in a mixture, we can mix different amounts of substances together and still have a mixture. But we can describe a mixture in two different ways depending on how the substances are mixed together. Let's learn about those ways today!

The first type of mixture is called a **heterogeneous mixture** (said this way: hĕt-ŭh-rå-jŭh-nŭhs mĭx-cher). Heterogeneous is a big word that simply means the substances in the mixture aren't evenly mixed together or distributed.

**materials needed**

- ☐ Hot water
- ☐ 2 clear glasses or coffee mugs
- ☐ 2 tea bags
- ☐ Sugar
- ☐ Olive or coconut oil
- ☐ Tablespoon
- ☐ Optional: milk
- ☐ Sticky tabs

Sometimes scientists use big words, but if we break those words down into parts and learn the meaning, it can help us understand and remember the word. In the word heterogeneous, "hetero" is a word that means different. This reminds us that the mixture is different or uneven.

You can think of a heterogeneous mixture kind of like a bowl of cereal. In the bowl, the cereal and milk are mixed together — but not evenly. Each bite you take will have different amounts of milk and cereal because the mixture is uneven.

Another example of a heterogeneous mixture would be orange juice with the pulp still in it. The pulp is mixed into the orange juice, but it is not evenly distributed through the whole glass. Sometimes, I take a sip of my orange juice and get a whole lot of pulp! What is the other type of mixture, Hannah?

The second type of mixture is called a **homogenous mixture** (said this way: hŭh-mŏj-ŭh-nŭhs mĭx-cher). In a homogenous mixture, the substances are evenly distributed throughout the whole mixture. We can break this word down too to help us remember it! "Homo" is a word that means the same.

Hmm, a homogenous mixture makes me think of frosting! All of the ingredients are mixed together evenly, and each bite tastes the same.

## Activity directions:

1. Ask your teacher to pour the hot water into both cups. Be sure to save space at the top of the cup for other ingredients.

2. Carefully steep the tea bags in the hot water according to the directions on the bag.
3. Discard the tea bags. Write "1" on a sticky tab and stick it to the table. Place one cup of tea above it. Write "2" on the other sticky tab and stick it to the table. Place the other cup of tea above the "2."
4. Add sugar and milk (optional) to each cup of tea.
5. Add 1 tablespoon of oil to the cup of tea above the "2" label.
6. We've created two mixtures! Examine each cup of tea. Which cup has the ingredients evenly distributed throughout the glass? Which cup of tea's ingredients are not evenly distributed throughout the glass? Answer the questions on the worksheet according to what you've observed.

1. Did cup of tea 1 or 2 have ingredients evenly mixed and distributed throughout it?

_______________________________________________

2. Would each sip of cup 1 taste the same?

_______________________________________________

3. Did cup of tea 1 or 2 have ingredients unevenly mixed and distributed throughout it?

_______________________________________________

4. Would each sip of cup 2 taste the same?

_______________________________________________

In a homogenous mixture, the substances are evenly distributed throughout the whole mixture. In a heterogeneous mixture, the substances are not evenly distributed throughout the whole mixture. Use your observations to label each cup of tea.

5. Cup of tea 1 was a _______________________ mixture.

6. Cup of tea 2 was a _______________________ mixture.

Welcome back! Today, we're going to learn about another type of mixture called a solution. But before we get started, we need to learn a new word. **Dissolve** (said this way: dĭh-zŏlv) is a word that means to melt or mix something into something else, like when we dissolve sweetened cocoa powder into milk to create chocolate milk.

You sure do like chocolate milk, Ben! It seems to be on your mind a lot lately.

**materials needed**

- Spoon
- Cup of cold water
- Cup of warm water
- Salt
- Teaspoon
- Stopwatch

I can't help it when we're talking about mixtures! Anyway, a **solution** (said this way: sŭh-loo-shŭhn) is formed when a substance is dissolved into a different substance. A solution is made of two parts: the solvent and the solute.

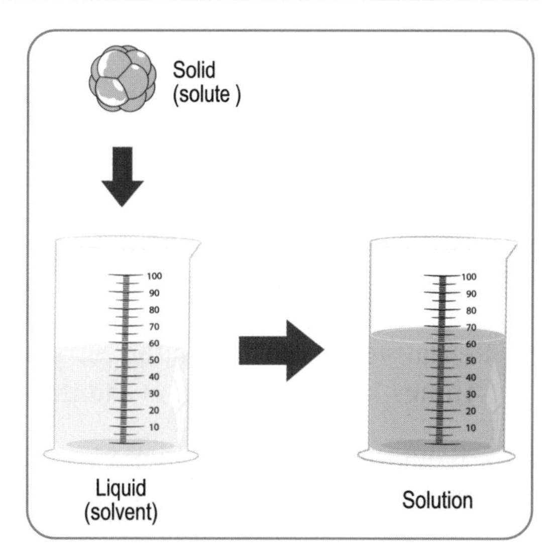


Let's start with the solvent! The **solvent** (said this way: sŏl-vŭhnt) is the substance that dissolves another substance. When it comes to chocolate milk, the milk would be the solvent. Water is another example of a common solvent.

On the other hand, the **solute** (said this way: sŏl-yoot) is the substance that is dissolved. So, we could dissolve sugar or salt into water — the sugar or salt would be the solute.

So the solvent is the substance that dissolves another substance, and the solute is the substance that is dissolved. Got it! I have a question, though — how is the solvent able to dissolve the solute?

Good question! The answer is the molecules in the solvent. The solvent's molecules work to break apart the molecules in the solute. Let's think about dissolving salt into water. The water is the solvent, and the salt is the solute. The water molecules push apart the salt molecules, which causes the salt to dissolve. Once the salt is dissolved, the water molecules keep them from joining back together.

Have you ever noticed it is easier to dissolve something when the solvent is warm? That is because when something is warm, its molecules move quickly. When something is cold, the molecules will move more slowly. When a solvent is warm, the molecules move quickly to separate the molecules of the solute. Let's give it a try!

## Activity directions:

1. It's time to create a lab report! Turn to the lab report worksheet on the next page and write your name and the date at the top. Then write down your question. Ben and Hannah's question is, "Will cold or hot water dissolve salt faster?"

2. Write your hypothesis, or what you think will happen. Ben and Hannah wrote, "The hot water's molecules move quickly and will dissolve the salt faster."
3. Ask your teacher or a friend to get the stopwatch ready. Ask them to start the stopwatch as soon as you add the salt and begin stirring. Then add a teaspoon of salt to the warm water and begin stirring. Once the salt is all the way dissolved, stop the stopwatch.
4. In the "Things I observed" section of the lab report, write "Warm water," then write down the time the stopwatch shows.
5. Repeat steps 3 and 4 with the cup of cold water. Once you're done, write "Cold water" in the "Things I observed" section and the time the stopwatch shows.
6. Fill out the Results section of your lab report. What happened in your experiment? Did the cold or the hot water dissolve the salt faster?
7. Bonus: How much salt can you dissolve in the warm water? Add 1 more teaspoon and stir to dissolve. Continue adding salt until the salt no longer completely dissolves. Once no more salt dissolves, we say that the solution has reached saturation. Once a solution has reached saturation, no more solute can be dissolved in the solvent.

HOT COLD

Name ______________________ Date ______________

# Lab Report

## Question

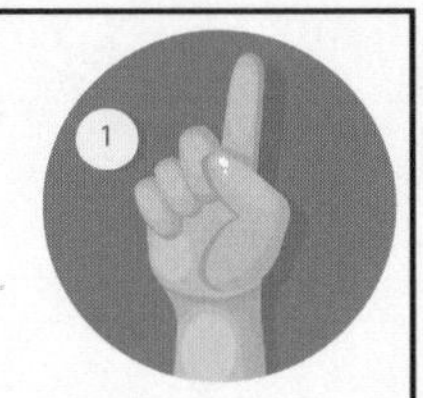

## Hypothesis

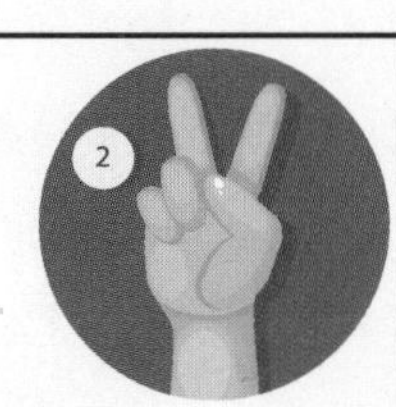

## Things I observed:

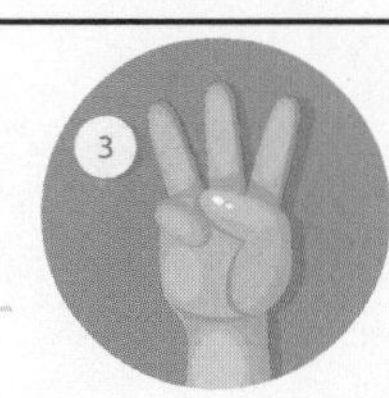

## Results

### What happened in the experiment?

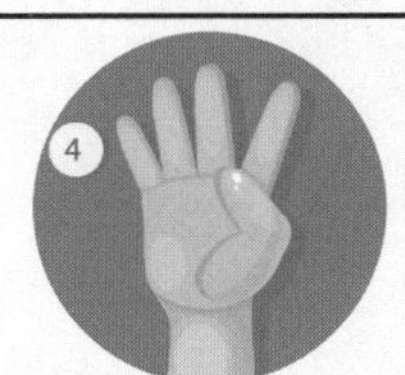

### Was my hypothesis correct?

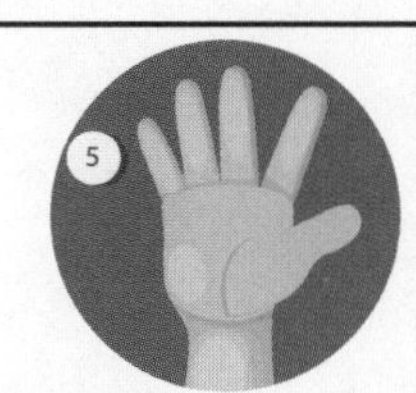

# Additional Lab Notes

I'm glad we were able to explore mixtures and solutions together this week. It was really interesting, and I had fun with our activities!

Me too! Learning about mixtures and solutions also reminded me of how our enemy lies. In the Bible, our enemy is called the serpent, the devil, or Satan. In John 8:44b, Jesus said this about the devil:

*He was a murderer from the beginning, not holding to the truth, for there is no truth in him. When he lies, he speaks his native language, for he is a liar and the father of lies.*

The devil is a liar, and he always has been. He is also crafty. We see at times in the Bible that he twists what is true or mixes just a little bit of truth into his lies. Let's read Genesis 3:1–5 together:

*Now the serpent was more crafty than any of the wild animals the LORD God had made. He said to the woman, "Did God really say, 'You must not eat from any tree in the garden'?"*

*The woman said to the serpent, "We may eat fruit from the trees in the garden, but God did say, 'You must not eat fruit from the tree that is in the middle of the garden, and you must not touch it, or you will die.' "*

*"You will not certainly die," the serpent said to the woman. "For God knows that when you eat from it your eyes will be opened, and you will be like God, knowing good and evil."*

The serpent was right that if Adam and Eve ate the fruit God had commanded them not to eat, then they would know good and evil. He mixed just a little bit of truth into the lie that they would not die as God had warned them.

Sometimes, lies are cleverly disguised and mixed in with the truth. In order to separate lies from the truth, we need to know the truth. John 14:6 tells us,

*Jesus answered, "I am the way and the truth and the life. No one comes to the Father except through me."*

In order to know the truth, we must have a relationship with Jesus and study the Word of God, the Bible. The Bible is our standard of truth — and it does not change. The better we know the Bible, the better we become at recognizing truth and lies in the world around us.

Talk with your family about ways you can learn more about the Bible together. Then look up John 14:6 in your Bible. If you'd like, you can highlight this verse in your Bible. Memorize John 14:6 with your teacher or with a sibling.

Oh good, you made it! I was afraid you might miss my favorite day of the week. It's time to add a new page to our Science Notebooks!

Ben has been excited to add his new page all morning — I think he needs to work on patience.

Maybe a little bit. But first, let's get to adding that new page! This week, we learned about mixtures and solutions. I was thinking we could draw an example of a homogenous and a heterogeneous mixture in our Notebooks.

Hmm, okay! We could draw a cup of tea or coffee or show a homogenous mixture and a bowl of cereal or orange juice or show a heterogeneous mixture. We don't have an example picture to use for these items, but I think we can use our imaginations and creativity.

We each chose to draw a cup of coffee and a glass of orange juice with some pulp. Here is how each of our drawings turned out. Have fun creating your Notebook page! We can't wait to see it.

In your Notebook, draw a picture of a cup of tea or coffee. Label this picture **Homogenous Mixture**.

Then draw a picture of a bowl of cereal or a glass of orange juice with pulp. Label this picture **Heterogeneous Mixture**.

Learning about mixtures and solutions this week also reminded us that lies can be mixed in with truth. Jesus is truth, and we learn about truth in the Bible. Copy John 14:6 on the back of your Notebook page as a reminder.

*Jesus answered, "I am the way and the truth and the life. No one comes to the Father except through me"* (John 14:6).

week 14

# Day 1 Chemistry Around Us

Welcome back! Are you ready to begin another science adventure together? I can't believe how much we've already learned — just think about everything we've explored so far: measurements, mass, matter, atoms, elements, molecules, the carbon cycle, and mixtures.

I've enjoyed each of our interesting science adventures together, but I'm also wondering what can chemistry be used for in everyday life? I mean atoms, molecules, and the elements are really interesting, but what can someone do with that knowledge?

That is a great question, Ben. Chemistry is actually an extremely useful field of science. Understanding atoms, molecules, and elements as the building blocks for all that we see helps us to better understand the living and nonliving things around us.

The scientists who study chemistry are called chemists. Chemists work to understand how atoms and molecules interact. They also work to understand the properties of different elements, substances, and compounds. Chemists can then share their knowledge with people who work in many other fields.

For example, let's say you're an inventor and you're working on the world's next great invention! But you have a problem: you need to find a metal that is a good conductor of electricity for your invention. If you talk to a chemist, they can tell you that copper and silver conduct electricity well. You may decide that silver is too expensive to use in your invention, but copper will work perfectly!

Inventors aren't the only people that can use chemistry to do their work. People who create or manufacture toys, tools, or equipment must also understand how elements, compounds, and substances will interact with each other. Dentists and doctors also use the field of chemistry to help them understand how things work together in the human body. Chemists and doctors can use their understanding of chemistry and biology to create treatments or medications to help someone.

- pH test kit or meter for soil 
- Local soil

Weekly materials list

Name: ____________________

Like me! I was coughing a lot, and sometimes it was hard to breathe. When Mom took me to see Dr. Jack, he discovered that I have asthma. Asthma is a disease that causes my airways and lungs to get tighter and smaller than they should be. That makes it hard to breathe! Dr. Jack gave me an inhaler with medicine in it that helps my lungs and airways work the way they should. I'm sure glad chemists and doctors can use their knowledge to help people!

Find each word in the word search:

chemist atom copper dentist doctor
inventor property manufacture

| | | | | | | | | | | |
|---|---|---|---|---|---|---|---|---|---|---|
| A | Z | Q | F | D | E | N | T | I | S | T |
| T | M | N | W | C | O | P | P | E | R | M |
| O | Q | Z | D | N | O | C | O | E | J | K |
| M | A | N | U | F | A | C | T | U | R | E |
| L | S | I | N | V | E | N | T | O | R | Z |
| P | R | O | P | E | R | T | Y | L | R | D |
| O | P | I | C | H | E | M | I | S | T | G |

Day 2

Hello there, friend! I'm excited to continue talking about how chemistry is used in the world around us today. Hey, Ben, you mentioned your asthma inhaler yesterday. Did you know the main medicine in your inhaler is made from atoms bonded together? The medication in your inhaler is called albuterol (said this way: ăl-byoo-tŭh-rŏl). Let's look at the chemical formula for albuterol: $C_{13}H_{21}NO_3$.

Hmm, let me check the periodic table of elements so that we can break that chemical formula down! Ah, this formula means that an albuterol molecule is made from 13 carbon atoms, 21 hydrogen atoms, 1 nitrogen atom, and 3 oxygen atoms all bonded together.

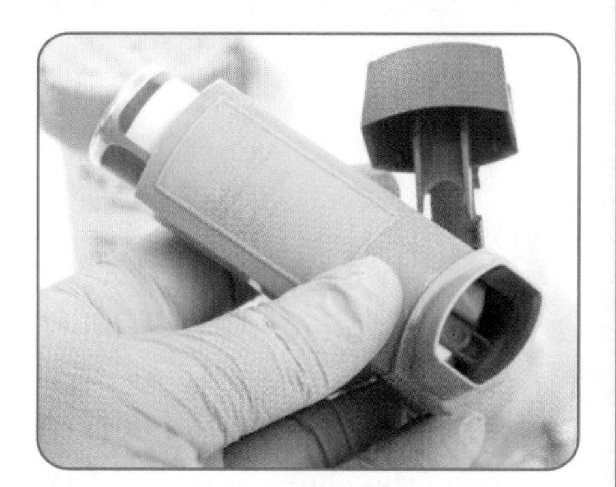

You also explained yesterday that asthma is a disease that causes someone's airways and lungs to get tighter and smaller than they should be. Doctors and biologists were able to use their knowledge and understanding to determine what was happening when someone had an asthma attack. The next step was to find something that would help!

Albuterol is a type of compound called a bronchodilator (said this way: brŏng-kō-dī-lā-ter). Bronchodilators work by relaxing the muscles that surround your airways. This helps you to be able to breathe air in the way you should be able to. Doctors and chemists were able to use their knowledge and understanding of elements and compounds to find a compound that would work to treat asthma.

Wow! So, scientists and doctors were able to work together to discover a disease, understand how the disease works in the human body, and find the right compound to help things work the way they should!

Exactly. Doctors and scientists often work together to understand different diseases and find compounds that can help someone's body work the way it was designed to. Without an understanding of chemistry, we wouldn't be able to understand the properties of elements and compounds — or how those properties may be able to help someone.

Doctors and dentists use many different medications and compounds to help people. Remember, an element's atoms combine in different amounts to create different compounds. Let's look at a few chemical formulas for some common medicines together.

Okay, this will be interesting! Before we get started, though, it's important to remember that we should never touch any type of medicine without an adult's permission and directions.

Name: ______________________________

apply it

1. Acetaminophen (said this way: ŭh-sē-tŭh-mĭn-ŭh-fŭhn) is commonly used to lower a high fever or reduce pain. Its chemical formula is $C_8H_9NO_2$. Use the periodic table of elements to find each element's symbol then write the element names below. Hint: N and O are symbols for different elements.

Acetaminophen is made from the elements:

______________________________

______________________________

2. Dentists often use procaine (said this way: prō-kān) to numb part of someone's mouth. Procaine allows a dentist to work on a tooth without the person feeling pain. The chemical formula for procaine is $C_{13}H_{20}N_2O_2$. Use the periodic table of elements to find each element's symbol then write the element names below.

Procaine is made from the elements:

______________________________

______________________________

3. Diphenhydramine (said this way: dī-fĕn-hī-drŭh-mēēn) helps reduce itching, fevers, or relieve allergies. The chemical formula for diphenhydramine is $C_{17}H_{21}NO$. Use the periodic table of elements to find each element's symbol then write the element names below. Hint: N and O are different elements.

Diphenhydramine is made from the elements:

______________________________

______________________________

4. Bonus! Did you notice that each of these medications is made from the same elements? What makes each medication different? Hint: Think about the numbers in each formula. What do the numbers stand for?

______________________________

______________________________

______________________________

______________________________

I'm glad we've been able to learn about some of the ways chemistry helps us in our everyday lives this week.

Me too — but we're not done yet with this week's adventure! Chemistry is a fascinating and helpful field of science for everything from inventing, creating, manufacturing, medication, and even farming.

**materials needed**

- pH test kit or meter for soil
- Local soil

Wait, farming too?

Indeed! Let's talk about how chemistry helps farmers. But first, we need to learn a new word. **Analyze** (said this way: ăn-l-līze) means to closely examine something. Through chemistry, farmers can analyze the soil to make sure that it has the nutrients their crops will need to grow well.

Ah, and that is important because different crops will require certain nutrients in the soil!

Right! Phosphorous (said this way: fŏs-fer-ŭhs), nitrogen, and potassium are all important elements that we find in the soil. If the soil is low in one of these elements, a farmer can fertilize the soil to help the crops grow. **Fertilize** (said this way: fur-tĭl-īz) means to apply something to the soil to add nutrients or make the soil richer. Farmers can use animal waste — which is also called manure — or certain chemicals to fertilize soil.

Another way farmers can analyze the soil before they plant a crop is to test the pH of the soil. This is a measurement of how much or how little acid is in the soil in a particular place.

Oh, Mom and I tested the pH of our soil before for her garden! Many substances around us can be classified as an acid or a base. You may remember at the beginning of our adventure this year that we talked about how vinegar is acidic. Acidic substances, like vinegar and lemon juice, taste sour and have a reaction with bases.

Bases have a bitter taste — like baking soda — and are often used as soaps and cleaning products. We can also call base substances alkaline (said this way: ăl-kŭh-līn). We measure pH using numbers. A substance with a pH measurement of less than 7 is classified as an acid. On the other hand, a substance with a pH higher than 7 is classified as a base. A substance with a pH measurement of 7 is said to be neutral — it is neither an acid nor a base.

Some types of plants prefer to grow in highly acidic soil, while other plants need alkaline soil. We can use a pH test kit or meter to test the soil where we live — let's give it a try!

## Activity directions:

ASK PARENT FOR HELP

1. Write your name and the date on your lab report on the next page. Next, write the question: Is our soil acidic or alkaline? Then create your hypothesis. What do you think your soil is?
2. Ask your teacher to help you gather some soil and test it according to your pH test kit or meter's directions. Determine the pH of your soil using the pH scale provided with your kit.
3. Write your observation in the "Things I observed" section of your report.
4. Finish your lab report and record the results of your experiment. Be sure to write down the pH of your soil in the results. Is your soil acidic, neutral, or alkaline?

Name ______________________ Date ______________

# Lab Report

**Question**

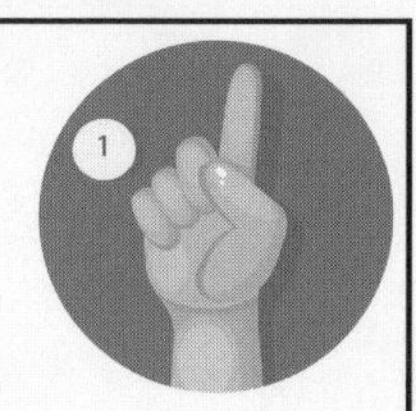


**Hypothesis**

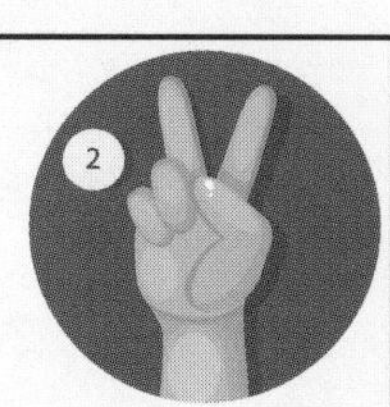


**Things I observed:**

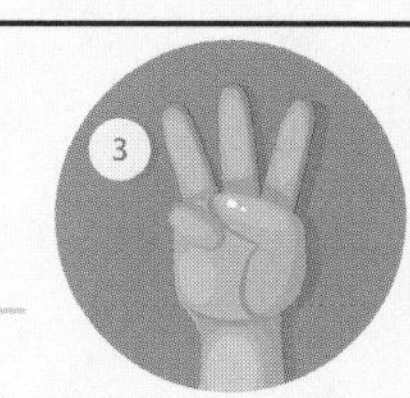


## Results

**What happened in the experiment?**

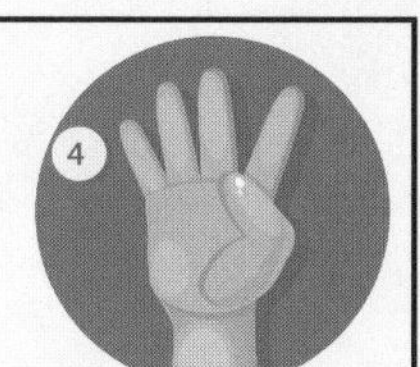


**Was my hypothesis correct?**

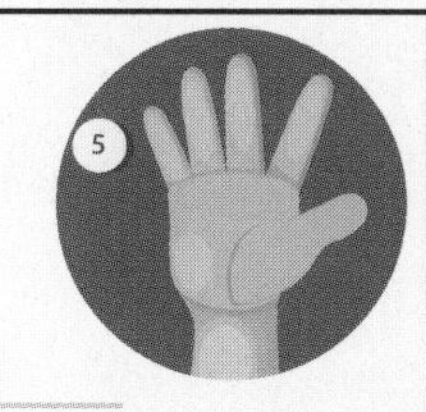

## Additional Lab Notes

I don't know about you, friend, but I sure had fun exploring the ways chemistry is used in the world around us! I thought it was especially interesting to learn a little more about the medicine that helps me to be able to breathe when my asthma is giving me trouble.

I'm so very grateful that God gave people curiosity and a desire to understand how things work. God also gave us many elements and compounds that can help our bodies when they aren't working quite right — I see God's merciful wisdom and provision for us in that!

You know, when things don't work quite right in our bodies, it reminds us that is because it was broken through sin. We learn in Genesis that God created a perfect world with no sickness or death. But when Adam and Eve chose to disobey God's direction and commit sin, it changed things. Sin breaks and destroys. Sin separates us from God, and it has consequences that affect everything from the plants, to the animals, to people, and to the whole world.

Because of sin's consequences, creation is no longer perfect, and we see sickness, sadness, and death. We see the effects of the broken creation all around us in the world and sometimes even in our own bodies, when things don't work together as they should. Though we still live with the effects of sin in the world, we also know that Jesus came to redeem us from our sin. Jesus came to earth to pay the price for sin on our behalf.

Through Jesus, we can have a relationship with God, and our sin is forgiven — it no longer separates us from God. We also have the promise that we will spend eternity with God in heaven. In the meantime, though, we still live in the creation that was broken by sin. Things may not always be perfect in our lives. . . .

Like when my asthma acts up, or when my favorite fish died, or when grandma had cancer.

But even when these things happen, we can still have peace because of Jesus and His promises to us. Peter was one of Jesus' disciples, and he wrote a letter to Christians who were suffering through very difficult things. Peter's letter is found in the Bible in the book of 1 Peter. In 1 Peter 5:10–11 he wrote,

*And the God of all grace, who called you to his eternal glory in Christ, after you have suffered a little while, will himself restore you and make you strong, firm and steadfast. To him be the power for ever and ever. Amen*

Just like we read in Revelation 21:3–4 a few weeks ago, we have hope in Jesus no matter what happens around us or what we may suffer. And someday, we know that He will restore things.

Look up 1 Peter 5:10–11 in your Bible. If you'd like, you can highlight these verses. Memorize 1 Peter 5:10–11 with your teacher or with a sibling.

Hey there, friend! Did you bring your Science Notebook with you today? I'm excited to add a new page.

Me too! I love being able to look back in my Notebook and see all the things we've learned about. This week, we talked about different ways chemistry is used in the world around us. I enjoyed talking about ways farmers can use chemistry when they plant their crops. Let's draw a picture of a farm field this week!

I like that idea. Here is a picture we could use for an example.

Here is how each of our farms turned out. I love how each one is unique and shows the creativity God gave each of us. Have fun with your drawing!

In your Notebook, write: Chemistry helps inventors, manufacturers, doctors, dentists, and farmers.

Then draw a picture of a farm field.

Learning about medications this week also reminded us that though things aren't always perfect in the world around us, we have hope in Jesus. Copy 1 Peter 5:10–11 on the back of your Notebook page as a reminder.

*And the God of all grace, who called you to his eternal glory in Christ, after you have suffered a little while, will himself restore you and make you strong, firm and steadfast. To him be the power for ever and ever. Amen* (1 Peter 5:10–11).

week 15

# Worldview I

Hello, and welcome back! Today we interrupt our regularly scheduled chemistry adventure to have an important conversation about the study of science.

What are you doing, Ben?

Oh, uh, I'm pretending to be a news reporter. I thought it would be a fun way to introduce our adventure for this week. May I continue?

You're always full of surprises, Ben! Go ahead.

Ahem, as I was saying. We've received reports from our sources —

You mean Mom and Dad?

Well, yeah, but don't interrupt! We've received reports from our sources that it would be helpful to take another look at the scientific method together. Let's first review our definition of science: Science is the pursuit of knowledge and understanding brought about through an organized process. Science helps us to ask questions, test our ideas, and share what we've learned with others. Through science, we also learn more about God and our relationship with Him.

You may remember that the organized process we use to study science is called the scientific method. We turn now to our reporter, Hannah. Hannah, would you please review the steps of the scientific method for us?

Sure thing — the scientific method helps us study science in an organized way as we ask questions, test our ideas, and develop conclusions. There are basically five steps to the scientific method.

1. Make an observation.
2. Ask questions.
3. Create a hypothesis.
4. Test it!
5. Share the results.

Back to you, Ben!

Sunglasses 

Family member

Weekly materials list

Name: ______________________________

Thanks, Hannah! These steps give us structure to study in an organized way. They are the method, or system, we use to study science. In our broadcasts, or uh, adventures, together so far, we've explored parts of chemistry like matter, atoms, elements, molecules, the carbon cycle, and mixtures. Scientists have discovered or learned more about all of these things by studying them through the scientific method. But the scientific method isn't the only thing that shapes how we study science.

We hope you'll join us this week for our investigative report on the scientific method! Over and out for now.

apply it

1. Fill in the blanks to complete the definition of science.

Science is the pursuit of ______________ and ______________ brought about through an ______________ ______________. Science helps us to ______________ ______________, test our ______________, and share what we've learned with ______________. Through science, we also learn more about ______________ and our ______________ with Him.

2. What other things do you think could shape the way we study science?

______________________________

______________________________

______________________________

______________________________

______________________________

______________________________

______________________________

______________________________

______________________________

Welcome back to this week's investigative report on science.

Um, are you still pretending to be a reporter, Ben?

Yes, Hannah. Please do not interrupt the broadcast.

Oh, okay. Sorry!

**materials needed**

- Sunglasses
- Family member

It has come to our attention that the scientific method gives us the organized structure we need to study science — but it isn't the only thing that shapes the way we approach studying science. In fact, our worldview has a large impact on how we see the world around us and how we study science. We turn now to our reporter, Hannah. Hannah, will you tell us what a worldview is?

Certainly! A **worldview** (said this way: wurld-vyoo) is what you believe and the way you see the world around you through your beliefs. Back to you, Ben!

Thank you, Hannah. Everyone has a set of beliefs that create their worldview — but not everyone's worldview is the same. To help us understand, let's turn back to our reporter, Hannah. Hannah, let's imagine that you and I are standing outside in the bright sunshine. Would we both have the same view of the world around us?

Yes, we would each see the same things around us.

Now, let's say that I put on a pair of sunglasses but you do not. How would that change the way we view things?

Hmm, well I would describe things as bright because of the sunshine. But your sunglasses would make the sunlight dimmer, so you would see and describe things differently. The sunglasses might also make colors look a little different to you compared to what I would see without the sunglasses.

Right! We would both see the same things around us — but we would describe things differently because our views are filtered differently through sunglasses or no sunglasses. Now, imagine that you have on a pair of glasses with blue lenses and I have on a pair with red lenses. What would happen?

Name: ____________________

We would still see all the same things in front of us, but we would see them very differently! I would believe everything is much more blue than it really is, and you would believe things are much more red.

In the same way, we all see the same things in the world around us. But just like the glasses would impact the way we see things, our worldviews impact what we believe about the world around us. That's it for our broadcast today, but be sure to join us for more news tomorrow!

## Activity directions:

1. Ask your teacher or a family member to help you with this activity. You may complete this activity inside or outside.
2. Choose who will wear the sunglasses first.
3. Take turns describing what you see. Describe the colors you see and whether the light is bright or not.
4. Now swap the sunglasses and describe what you see again. Write down your observations from this activity on the worksheet below.

apply it

1. What did you both see in the world around you?

____________________

____________________

____________________

2. Did you both see things the same way, or did the sunglasses change things?

____________________

____________________

____________________

3. How did what you see change when you swapped the sunglasses?

____________________

____________________

____________________

Reporter Ben here, and we're back with another live report as we investigate worldviews this week. In our previous broadcast, we learned that a worldview is what you believe and the way you see the world around you through your beliefs. Today, we're going to explore our worldview as Christians. Let's turn again to our reporter, Hannah. Hannah, what shapes our worldview as Christians?

Excellent question, Ben. As Christians, our worldview is shaped by the Bible. We turn to the Bible first to tell us what truth is, to learn how we are to live our lives before God, and to reveal how we can have a relationship with God. We believe what is written in the Bible. In other words, the Bible is the foundation for the way we view the world around us. Just like a pair of sunglasses impacts what we see, the Bible impacts our beliefs, how we view the world, and everything we study.

In fact, the Bible shapes the way we study science, math, history, and even language! For example, as we learn about math, we discover that math is consistent. Two plus two always equals four. But why is that? Well, math is consistent because God is consistent. We can see order, logic, and consistency in math because these are the attributes of God — and God created math.

Interesting! What about language? How can the Bible shape the way we view language, Hannah?

We see in Genesis that God spoke and created all that we see and even the things that we cannot see, like atoms. Throughout the Bible, we see that God has also communicated with mankind. God communicates through language — and He gave us the ability to communicate as well. The Bible also reveals why we have different languages!

Ah, in Genesis 11:1–9! Our view of history is also shaped by the Bible. In the Bible, we see that history is the unveiling of God's plan of salvation for mankind and the eventual restoration of creation. No matter what we study in history, we view it through God's ultimate control over all things.

When the Bible is our foundation, it also becomes the foundation for the way we explore everything around us. It shapes the way we see things — and how we interpret the evidence we find in science. We'll be talking more about that next week. Be sure to join us then!

Name: ______________________________

1. What is a worldview?

2. What is the foundation of our worldview as Christians?

3. Read Genesis 11:1–9 in your Bible. Why do we have different languages in the world?

I had fun learning about worldviews this week, and I'm excited to talk more about them next week. In the meantime, I learned that my worldview is shaped by the Bible. What I've learned about God from the Bible impacts how I see everything in the world around me.

Me too! As I learn more about God's creation, I see His handiwork. It reminds me of what David wrote in Psalm 19:1–4,

*The heavens declare the glory of God; the skies proclaim the work of his hands. Day after day they pour forth speech; night after night they reveal knowledge. They have no speech, they use no words; no sound is heard from them. Yet their voice goes out into all the earth, their words to the ends of the world.*

Do you remember when we went on our camping trip this summer, Hannah?

Of course I do! We laid out underneath the night sky together and tried to count all the stars that we could. There were so many stars — they were impossible to fully count!

Dad read Psalm 19:1–4 to us once we were done counting. Then he explained that like the verses say, the sun, moon, and stars can't actually speak to us with words — but they are a display of God's glory. They show the glory of God throughout the whole earth.

David's worldview — remember, he is the one who wrote those verses — was shaped by what he had read in the Scriptures and by what he had experienced through God in his life. Because of his worldview, he saw the stars as more than simply bright lights in the sky. When the Bible is our foundation, we know that God created all that we see. The stars are His creation and are a display of His glory, power, and majesty.

The stars are one of my favorite reminders of God's glory. Let's count as many as we can tonight!

The next time you have a clear view of the night sky, see how many stars you can count. Look up Psalm 19:1–4 in your Bible. If you'd like, you can highlight these verses. Memorize Psalm 19:1–4 with your teacher or with a sibling.

We interrupt our special worldview report to bring you Science Notebook day!

Hey, you took my lines! But I guess I'm not upset — I'm too excited to add a new page to my Science Notebook!

Oh good, you made reporting look like so much fun that I wanted to give it a try too. This week, we talked about our worldview and how it impacts the way we see the world. I was thinking we could draw a set of glasses in our Notebook this week.

Ooh, I like that idea. We could draw a pair of sunglasses — here is an example we can use.

Wait, how about you and Sam draw a picture of glasses, Hannah — I'm going to draw a picture of a monocle on mine. We'll be sure to show you how each of our Notebooks turned out. Have fun with your drawing!

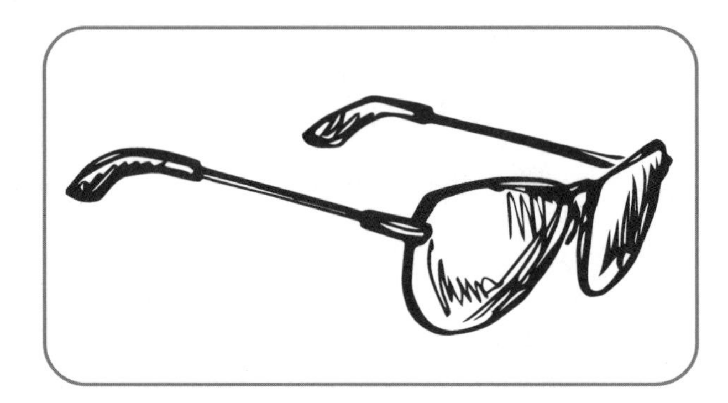

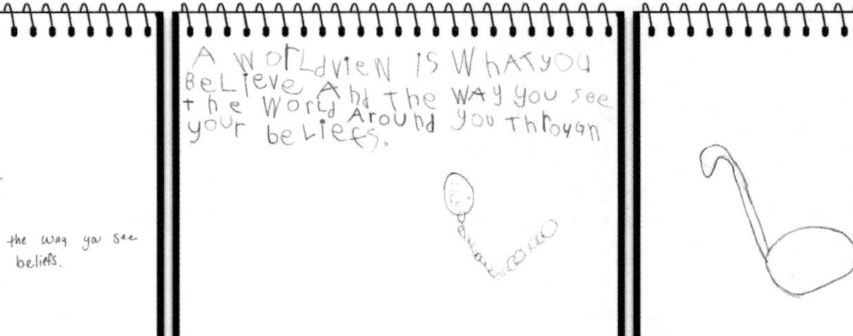

In your Notebook, write: A worldview is what you believe and the way you see the world around you through your beliefs.

Then draw a picture of glasses.

Learning about worldview this week also reminded us that the Bible is the foundation of our worldview as Christians. Copy Psalm 19:1 on the back of your Notebook page as a reminder.

*The heavens declare the glory of God; the skies proclaim the work of his hands* (Psalm 19:1).

# Day 1 Worldview 2

Hello again! We're going to continue our special exploration of worldview this week. I'm glad you're back to learn more with us! Last week, we learned that our worldview is what we believe and the way we see the world around us through our beliefs.

As Christians, we believe that the Bible is God's Word to us and that it reveals truth. Ben, I understand how the Bible is the foundation for the way we see everything in the world around us. But can you explain how the Bible impacts the way we study science?

I thought you would never ask — that's exactly what we're going to dive into this week. Let's start with evidence. As we study science, we see all kinds of evidence. **Evidence** (said this way: ĕv-ĭ-dŭhns) is the information we can see or experience around us. We may see evidence when we observe something in nature, examine a fossil, or do an experiment.

Like when we tested the reaction of lemon juice and baking soda. When we combined them, we saw bubbles and foam. The bubbles and foam were the evidence that there was a chemical reaction between those two things.

Right! Do you remember when we learned about Mendeleev's work to organize the elements? His work with the elements revealed evidence that there were elements that hadn't been discovered yet. A biologist studying plants may uncover evidence of the process of photosynthesis. Or a geologist studying rocks may find evidence that particular rocks were formed by water.

Sometimes, the evidence we see is very clear — like the chemical reaction of lemon juice and baking soda. Other times, however, the evidence isn't quite so clear about what happened or what is happening.

For example, a geologist may find evidence that certain rocks were formed by water. In this case, the evidence points to the rocks being formed by water — but the evidence alone can't answer all of our questions, such as when did the rocks form? Or what caused the water to form these certain rocks?

Name: ______________________

In these cases, scientists may look for additional clues to help interpret the evidence. **Interpret** (said this way: ĭn-tur-prĭt) means to explain what something means or to develop an understanding. What we believe about the world — our worldview — has an impact on the way we interpret the clues and evidence we see in science. We're going to talk more about that this week. But first, let's see how we might interpret some evidence!

apply it

1. Imagine you left a cookie on the countertop. When you return, you find that your cookie is gone! The evidence is that your cookie is no longer where you put it. What do you think happened to the cookie?

2. What you think happened to the cookie is your interpretation of the evidence. Why did you interpret the evidence this way?

3. Your interpretation of the evidence may be that you may think the cookie fell off the countertop, that someone took it, or even that someone else ate your cookie! How might you prove your interpretation of the evidence?

Welcome back! How did you interpret the evidence of the missing cookie, friend?

I can tell you that Hannah blamed me for her missing cookie. Can you believe that?

Well, that was my interpretation of the evidence based on the fact that you've accidentally eaten my cookie before!

Okay, well, that is true — sorry!

That's okay! But in the case of the missing cookie, I assumed that since you've eaten my cookie in the past, it was likely that you would eat it again now. It just made sense!

Ooh, that reminds me of the topic we're going to explore today. When scientists interpret evidence, they create a logical explanation for the evidence. **Logical** (said this way: lŏj-ĭ-kŭhl) means something that would be expected or makes the most sense.

In the case of the missing cookie, Hannah knew that I had accidentally eaten her cookie in the past. That's a fact! Because of that, it would make sense that I could have eaten her cookie this time too. She created a logical explanation for what happened to the missing cookie based on the evidence and facts.

In the same way, scientists must examine evidence and facts to create logical explanations for what we observe, find, or see in the world around us.

It all starts with the scientific method. First, a scientist makes an observation and asks questions. Then the scientist can create a hypothesis and test it. The result of the test or experiment provides evidence that may support or disprove the scientist's hypothesis.

When there is evidence to support a hypothesis, the hypothesis can become a scientific **theory** (said this way: thē-ŭh-rē). A scientific theory is a logical way to explain what we see or to answer a question based on evidence and facts. Once a theory has been developed, a scientist can continue working to support the theory with additional evidence.

It's similar to the case of the missing cookie. For example, I could have asked Ben if he had eaten the cookie or asked if anyone else had seen Ben eat the cookie. This would have given me additional evidence to support or disprove my theory.

Yup! As we learn about scientific theory, it's important to know that just like Hannah's theory was based on what she has experienced and believed to be true, a theory in science can be affected by what the scientist has experienced or believes to be true.

In other words, our worldview has an effect on how we might interpret evidence or explain something?

Name: ____________________

Correct! We call this a bias (said this way: bī-ŭhs), and we're going to talk more about it tomorrow.

Copy each definition below.

1. Logical means something that would be expected or makes the most sense.

2. A scientific theory is a logical way to explain what we see or to answer a question based on evidence and facts.

Are you ready for today's science adventure? Buckle up, we have quite a ride in store for us today!

So far in our adventures together, we've explored the field of chemistry. But soon, it will be time to turn our focus and begin exploring biology. Biology is the field of science that studies living things, and you may remember that all living things are made from atoms bonded together! Isn't it cool how chemistry is also related to biology?

Scientists who study biology are called biologists. Biologists ask big questions like: How does a living thing behave? What does a living thing need to survive? Where did the living thing come from in the beginning?

Hmm, that last question is different than the first two questions. We can observe and test how living things behave and what they need to survive, but we can't observe and test how a living thing came to be for the very first time.

That is true. For a question like this, a biologist creates a hypothesis to test. If there is evidence to support the hypothesis, then the explanation becomes a theory. Remember, a scientific theory is a logical way to explain what we see or to answer a question based on evidence and facts. However, theories are often affected by our worldview.

That brings us back to where we left off yesterday. Our worldview affects the way we interpret evidence. This is called a **bias** (said this way: bī-ŭhs). A bias is a belief, opinion, or worldview that shapes how we see and interpret the world. Everyone has a worldview bias, and it is the starting point for how we interpret evidence.

As we explore science, we find that there are basically two different starting points, or worldviews. In the first worldview, a scientist believes that God created the heavens and the earth just as the Bible tells us. This worldview is often called a biblical worldview.

In the second worldview, a scientist believes that the world wasn't created — all that we see just happened and came to be over millions of years through evolution. This worldview can be called a secular or evolutionary worldview.

Ah, I can see how a bias or worldview would impact the way we interpret evidence in science, then. As Christians, our worldview is based on the Bible. The Bible tells us that God created the heavens and the earth. Our worldview will affect the way we study science because we'll see God's handiwork on display in what we study!

But in a secular worldview, the evidence would be interpreted without God and without what the Bible tells us.

**Name:** ____________________

Right. Because the evidence is interpreted from two very different starting points, the conclusions about the evidence are also very different. We're going to be talking about this more as we continue our science adventure in biology together.

apply it

1. What is a bias?

____________________

____________________

____________________

____________________

2. What do you believe about the Bible?

____________________

____________________

____________________

____________________

Whew, what a week it has been! We've been exploring how our worldview bias affects the way we interpret evidence in science. In one of our very first adventures together, we talked about how science is the pursuit of knowledge and understanding. Science can help us understand more, but it cannot give us complete truth — only God and His Word can do that.

But why can we trust the Bible?

I talked with Mom about that just last night before bed! We read in 2 Timothy 3:16–17 that

*All Scripture is God-breathed and is useful for teaching, rebuking, correcting and training in righteousness, so that the servant of God may be thoroughly equipped for every good work.*

One of the best ways to determine if we can trust something or not is to look at the evidence we see. The Bible records many, many different historical events and places. **Archaeology** (said this way: årk-ē-ŏl-ŭh-jē) is a field of science that studies historical events. Archaeologists work to find evidence from historical events and places — and what they find confirms just what the Bible describes.

Another piece of evidence is that in the Bible, we read about many different prophecies. A prophecy is a prediction about what will happen at a future time. The Bible has hundreds of prophecies that accurately described what would happen many, many years before it actually happened. Many of the prophecies told about the coming of Jesus — and Jesus fulfilled each and every prophecy about Him.

We also find evidence through science that lines up with what the Bible says about scientific topics and events, like the worldwide Flood of Noah's time. The Bible is a very unique book that has stayed consistent and accurate throughout history. These are just a few of the things we can look at to give us evidence that the Bible is God's true and trustworthy Word to us.

Thanks for sharing, Hannah! I definitely want to learn more about why we can trust the Bible — let's talk to Mom and Dad about it at dinner tonight.

In *The Answers Book for Kids Volume 3,* read the sections "Where Did the Bible Come From?", "Why is the Bible True?", and "Why Do People Believe Different Things?" Discuss these sections with your family (this activity is optional). Look up 2 Timothy 3:16–17 in your Bible. If you'd like, you can highlight these verses. Memorize 2 Timothy 3:16–17 with your teacher or with a sibling.

I don't know about you, but I sure am ready to add a new page to my Science Notebook this week!

I've got our art supplies right here and ready to go. This week, we explored how our worldview impacts the way we interpret evidence in science. As Christians, our worldview is based on the truth of God's Word in the Bible. Hey, that gives me an idea!

Hmm, I think I know what it is. Let's draw a picture of a Bible in our Science Notebook this week.

Yes! We can use our own Bible as the example for our image. Are you ready to get started? Let's go!

Here is how each of our Bible drawings turned out.

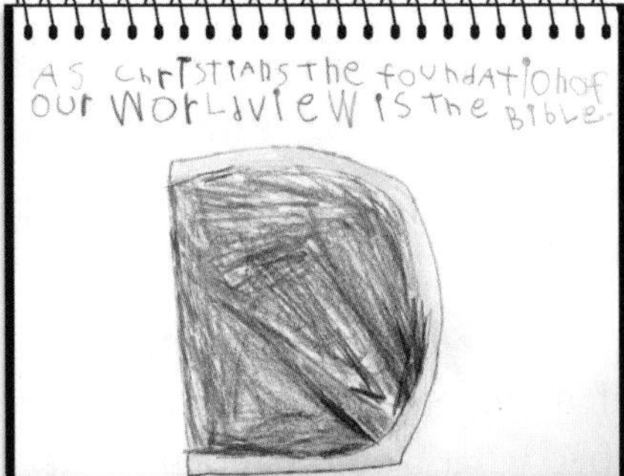

In your Notebook, write: As Christians, the foundation of our worldview is the Bible.

Then draw a picture of your Bible.

Learning about our worldview this week also reminded us that we can trust the Bible to be our foundation. Copy 2 Timothy 3:16–17 on the back of your Notebook page as a reminder.

*All Scripture is God-breathed and is useful for teaching, rebuking, correcting and training in righteousness, so that the servant of God may be thoroughly equipped for every good work* (2 Timothy 3:16–17).

week 17

# Observational & Historical Science

Day

Hey there, you're just in time to get started on our science adventure for this week.

We've been learning what a worldview is and how it affects the way we interpret evidence in science. Understanding someone's worldview is going to be important as we begin to explore the field of biology together next week. But before that, we need to talk a little more about two types of science that we encounter in the field of biology. Are you ready to dive in?

Let's start with observational science! You may have guessed that observational science has something to do with the things we can observe — and you'd be correct! **Observe** (said this way: ŭhb-zerv) means to see or notice something. Observational science is science that we can see, experience, or observe. This type of science is observable, testable, and repeatable. Let's think of an example.

I've got one! We learned about acids and bases a few weeks ago when we measured the pH of our soil. We can test and observe that vinegar is an acid and baking soda is a base. We can test them to see if they will react with each other. We can also repeat this experiment many times to see that each time we combine them, there is a reaction.

In observational science, we can test things and show others the results of our experiments. Other people can then repeat our experiments to see if their results are the same or different. Remember, observational science is observable, testable, and repeatable.

Observational science can also be called empirical science. **Empirical** (said this way: ĕm-pir-ĭ-kŭhl) means that something can be supported or verified through an experience or an experiment.

Biology often explores empirical science through what we observe in creation. For example, if a biologist decides to study rabbits, they can observe a rabbit's diet. A diet is what a person or animal tends to eat. The biologist can also observe where rabbits live, their size, and their behavior from day to day.

Name: ______________________________

Over time, the biologist can begin to describe a rabbit's habitat, size, diet, and behavior. The biologist can even test and observe rabbits in other areas to see if they are similar or different. Other scientists may also observe the rabbits to verify that the data collected is accurate. In other words, the biologist has collected information or data on rabbits that can be observed, tested, and repeated through more observation. This is how observational, or empirical, science works!

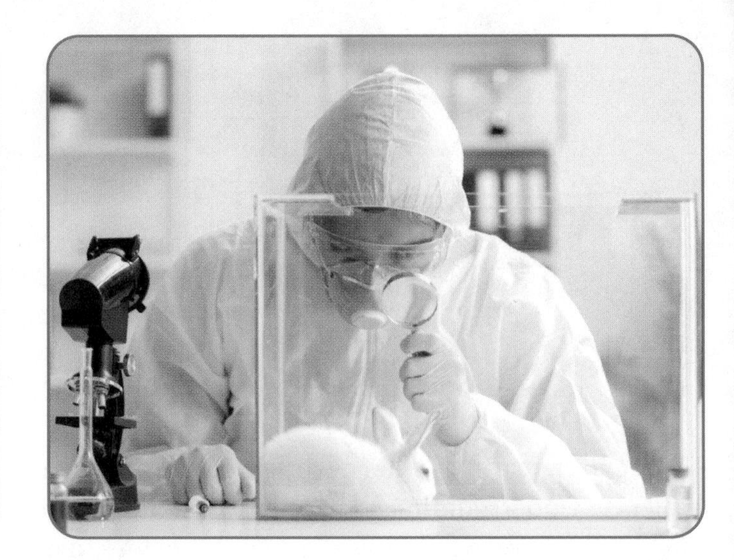

Fill in the blanks to finish the sentences.

1. Observational science is ______________, ______________, and ______________.

2. Empirical means that something can be ______________ or ______________ through an ______________ or ______________.

3. What is something you would like to study or observe?

______________________________

______________________________

______________________________

______________________________

4. How would you study or observe this like a scientist?

______________________________

______________________________

______________________________

______________________________

We talked about observational science yesterday. Can you remind us what we learned, Ben?

Sure thing. Observational science is science that we can see, experience, or observe. This type of science is observable, testable, and repeatable. We can also call observational science empirical science. Remember, empirical means that something can be supported or verified through an experience or an experiment.

You have a great memory, Ben! The second type of science that we sometimes encounter in the field of biology is called historical science.

Hmm, from that name, I would have to guess that this type of science deals with the past?

Correct. As we explore creation through science, we ask many different questions. Some questions are easy to answer, like how much does a rabbit usually weigh? To answer that question, we could observe and weigh many rabbits to find out what a rabbit usually weighs. We can then share what we've learned about a rabbit's typical weight. This would be observational science — it is observable, testable, and repeatable.

Got it! Sometimes, though, we may ask questions that aren't as easy to answer. What if our question was, how long have rabbits been on the earth?

That is a tricky question because we cannot observe how long rabbits have been on the earth. We can't test how long rabbits have been on the earth. And if we can't test it, we certainly cannot repeat it either! Our answer to this question is part of historical science — and it is based on our worldview.

Ben, how would we start to answer this question from a biblical worldview?

Well, we would want to start with God's Word — the Bible! Let me think . . . a rabbit is a living creature, and it lives on land. According to Genesis 1:24–31, God would have created rabbits on the sixth day of creation. But that still doesn't tell us exactly how long rabbits have been on the earth.

Name:

Good point, Ben. We can't observe, test, or repeat how long rabbits have been on the earth. Since we cannot observe, test, or repeat an experience or experiment to answer this question, we know we are trying to answer a question about historical science.

Historical science cannot be observed, tested, or repeated. To answer our questions about historical science, we can only create a hypothesis. If there is evidence to support our hypothesis, then it becomes a scientific theory. Mom is calling us for lunch now — but we'll talking about this tomorrow!

apply it

1. Read Genesis 1:24–31 in your Bible. What did God create on the sixth day of creation?

2. What did God give to living things for food after He created them?

3. How are historical and observational science different?

Oh good, you're here! I'm excited to begin our science adventure today. I've been puzzled since yesterday about how we begin to create a theory to answer a question about historical science. Since we can't observe, test, or repeat the past, how can we begin to create a hypothesis and theory?

Great question, Ben. Remember, a scientific theory is a logical way to explain what we see or to answer a question based on evidence and facts. As we begin to answer a question about historical science, we develop a hypothesis. Then we can look at evidence and facts to see if they support or disprove our hypothesis.

Historical science isn't testable, observable, or repeatable, however. How would we look for evidence to support our hypothesis?

We may look at evidence that is available through fossils, records, or eyewitness accounts. An eyewitness account is something that someone once documented in history.

Like a picture someone drew in a cave or on pottery, or something someone once wrote down.

Right. So, back to our question from yesterday: how long have rabbits been on the earth? As we gather evidence, we might look for the very first time someone wrote about a rabbit in history. If we know when this was written, it can help to give us a date or a time frame to start with. In other words, it gives us evidence that rabbits were on the earth and a time in history that we can work with. But this still is not the definite answer to our question because we can't observe, test, or repeat history.

Interesting — so we can use information and evidence to help us develop a hypothesis and then a theory in answer to a historical science question, but we won't be able to prove our theory.

Exactly.

I have a question now. I love to read, and I read a lot of books about science. How do I know when I'm reading about observational or historical science? How can I tell what can be proven and what is a theory?

Excellent question. If you're not sure, the best question to ask is, "Can this information be observed, tested, and repeated?" If it can be, you're reading about observational science. If it cannot be, you're reading a theory about historical science.

Let's practice! Is this observational or historical science? "Bottlenose dolphins can grow to be between 6 and almost 13 feet long."

Name: ____________________

That would be observational science because we can observe bottlenose dolphins, test their lengths, and repeat our measurements by measuring other dolphins.

You got it! Let's look at some more examples together.

1. What question can you ask yourself to tell if you're reading about historical or observational science?

____________________

____________________

____________________

____________________

Read each sentence then circle historical science or observational science. Don't forget to ask yourself, "Can this information be observed, tested, and repeated?"

2. Vinegar and baking soda have a chemical reaction when they are mixed together.

   **Observational Science / Historical Science**

3. Rabbits have been on the earth for about 6,000 years.

   **Observational Science / Historical Science**

4. A tiger can run between 30 and 40 miles per hour.

   **Observational Science / Historical Science**

Hello! We've learned a lot about observational and historical science this week, haven't we? I have a feeling this information will be helpful as we explore the field of biology together.

It's important to remember that the Bible is the foundation we use as Christians to help us interpret evidence as we explore creation and develop theories to answer questions. The Bible is the foundation of our worldview — and it is also an eyewitness account!

God was there at the very beginning; He knows exactly how the world came to be. God inspired men like Moses, the prophets, Matthew, Mark, Luke, John, and Paul to write His words through the Holy Spirit. In 2 Peter 1:21 it says,

*Prophecy never came simply because a prophet wanted it to. Instead, the Holy Spirit guided the prophets as they spoke. So, although prophets are human, prophecy comes from God* (NIrV).

Exodus 24:4 also says,

*Moses then wrote down everything the LORD had said.*

Did you know that the Bible gives us a timeline of history as an eyewitness account? A timeline is a way we can record dates and events to help us see when they happened in history.

We find genealogies recorded in the Bible. A genealogy is the record of a family or group. We can find an example of a genealogy in Genesis chapter 5. This genealogy tells us about Adam's family line — and it also tells us how long they lived.

We can use the information in the Bible's genealogies to create a timeline that shows us how long each person lived. Men and women have added up the ages found in the Bible, and they discovered that the earth is around 6,000 years old.

So, we can use the evidence we find in the Bible to create our hypothesis for how long rabbits have been on the earth. Using the Bible's timeline and the evidence we find to support it, our theory would be that rabbits were created on the sixth day of creation and have been on the earth for around 6,000 years.

Our theory is based on the evidence we find and our worldview. Our worldview is that we believe the Bible gives us an eyewitness account of history from the very beginning. We believe that the Bible is true and trustworthy, and it is the foundation for how we view the world.

In *The Answers Book for Kids Volume 3,* read the sections "Who Wrote the Bible?" and "How Do We Know What Really Happened?" You can also look at a timeline based on the genealogies found in the Bible in the *Big Book of History* (these activities are optional). Look up 2 Peter 1:21 in your Bible. If you'd like, you can highlight this verse. Memorize 2 Peter 1:21 with your teacher or with a sibling.

Why, hello there! Do you know what day it is?

It's time to add a new page to our Science Notebook! We explored observational and historical science this week, and it was really interesting. I have an idea for what we can add to our Science Notebook.

Okay, I have our art supplies ready. What do you think we should draw?

Let's write down what observational science is — then draw a picture of something that we can observe. One of our activities mentioned a bottlenose dolphin. That would be fun to draw!

Ooh, that does sound like fun. We could even write down how long they usually grow to be. Let's get started. Here is a picture we can use for an example.

Here is how each of our drawings look. Isn't Sam's dolphin cute? I think my dolphin looks more like beluga whale — but I still like it. Have fun creating yours!

In your Notebook, write: Observational science is observable, testable, and repeatable.

Then draw a picture of a bottlenose dolphin. Bottlenose dolphins can grow to be between 6 and almost 13 feet long. You can write down the length you think your dolphin would be.

Learning about observational and historical science this week reminded us that the Bible is an eyewitness account. God inspired men to write His words through the Holy Spirit. Write 2 Peter 1:21 on the back of your Notebook page as a reminder.

*Prophecy never came simply because a prophet wanted it to. Instead, the Holy Spirit guided the prophets as they spoke. So, although prophets are human, prophecy comes from God* (2 Peter 1:21; NIrV).

week 18

# Living Things

Hello! Are you ready for our next science adventure? We've explored many parts of chemistry together, and now it's time to turn our focus to the field of biology.

Woohoo! I'm excited to begin. Biology is a field of science that studies living things, and there are so many different living things to learn about. Just think about it — a biologist can study the human body, animals, flowers, trees, plants, ocean life, cells, ecosystems, and more!

Biology is definitely a broad field of science. There is so much we could explore — but let's start by talking about living things.

All right! My first question is, where did living things come from?

That is a great question, Ben. It is also one of the historical science questions that many biologists ask. We cannot observe, test, or repeat where the first living things came from — but we can look for eyewitness accounts and evidence to help us support our theory.

Then we should begin with the Bible because it is the foundation of our worldview. God was there at the very beginning, and He knows how life came to be!

Yes, He does! To form our worldview, we begin with Genesis 1:1,

*In the beginning God created the heavens and the earth.*

Whoa, hold on a second. I just remembered we learned before that there are basically two different worldviews. The biblical worldview, which is our foundation, begins with God and how He created the world as the Bible tells us. But what does a secular, or evolutionary, worldview begin with?

Well, the evolutionary worldview does not begin with God or with His Word. That means that this worldview develops very different theories to answer questions about historical science. In the evolutionary worldview, life and the universe just happened over the course of millions and billions of years. This is called the theory of evolution.

Name: ____________________

You may have heard parts of the theory of evolution as you read books or watched television shows about science.

Hmm, if we're going to spot the differences between these worldviews, then we need to make sure we understand our worldview really well. Let's go back to Genesis to learn what the Bible has to say about where living things came from.

apply it

Read Genesis 1 in your Bible. What did God create on each day? Write your answer.

1. On the first day, God created ____________________.
2. On the second day, God created ____________________.
3. On the third day, God created ____________________.
4. On the fourth day, God created ____________________.
5. On the fifth day, God created ____________________.
6. On the sixth day, God created ____________________.
7. Now read Genesis 2:1–3 in your Bible. What happened on the seventh day?

____________________

____________________

____________________

____________________

8. According to the Bible, where did living things come from?

____________________

____________________

____________________

____________________

Welcome back! Yesterday, we developed our biblical worldview. In order to answer the historical science question, "Where did living things come from?" we went straight to the Bible to read an eyewitness account. In the Bible, we learned that God created the heavens, the earth, and all living things in six days. Then He rested on the seventh day. This is our worldview foundation as we answer questions about historical science.

Our question, "Where did life come from?" is a historical science question. To answer this question, we start with the Bible to form our hypothesis. Based on what we read in the Bible, our hypothesis is that God created the heavens, the earth, and all living things just as it says in Genesis. Scientists with a biblical worldview examine the evidence found in the world around us to support this hypothesis. Their theory can be called **creationism** (said this way: krē-ā-shŭh-nĭz-ŭhm).

Now, because this theory answers a question about historical science, we cannot observe, test, or repeat history to prove our theory the way we could with observational science. But we can look for additional evidence that supports our theory.

We've talked before about how science is all about organizing things. Why do you think we are able to organize things in science so well, Ben?

Well, according to our worldview, God is the Creator. We learn in 1 Corinthians 14:33 that God is not disorganized; He is a God of organization and peace. So, we would expect to also see organization in His creation.

Indeed. We find organization all through creation. One way scientists can organize things is by separating creation into groups of living and nonliving things. Living things have certain traits or properties that can help us identify them. Let's think of some things that living things must be able to do.

Well, to begin, a living thing must be able to breathe. Humans and animals breathe air through nostrils — that's a big word for the holes in our nose. Birds breathe through nostrils in their beaks, and fish breathe through their gills.

Name: ______________________________

Another word that we can use for breathe is respiration. **Respiration** (said this way: rěs-pŭh-rāy-shŭn) is the process of breathing. Usually, living things inhale or absorb oxygen and exhale or release carbon dioxide through respiration. Hmm, plants are also grouped into the category of living things. Can you tell us how plants breathe, Ben?

Sure thing. Plants breathe through the process of photosynthesis as they absorb carbon dioxide and release oxygen.

Excellent! There are a few more traits that we can use to identify living things — but we're out of time for today. We'll pick this adventure up again tomorrow.

1. Respiration is one way we can identify living things. Based on this information, how many living things can you find in or around your home?

2. What other ways do you think we might be able identify living things from nonliving things?

Hello, friend. We were just about to continue our science adventure from yesterday. Remember, living things can be identified by certain traits or properties that they all have in common. For instance, all living things must breathe.

Another way we can identify living things is that they can grow. A baby grows into an adult, a puppy grows into a dog, a baby bird will become an adult bird, fish grow bigger, and plants grow from a small seed into a larger plant.

Wait, what was that noise?

Oh, just my stomach growling. I'm awfully hungry and ready for lunch!

Speaking of lunch, living things also need food to provide nutrients.

Ah, that makes sense! **Nutrients** (said this way: new-trēē-ĕnts) are a substance that plants, animals, and people need to grow and live. Nutrients provide the energy that living things need. People, birds, animals, and fish absorb nutrients from the food they eat. Plants absorb the nutrients they need through their roots.

And speaking of energy, living things can also move on their own, or respond to the environment around them. I can move my body, and so can animals, birds, and fish. But what about plants?

Ah, plants are interesting living things indeed! While they don't have a body or limbs like people, animals, birds, and fish, they can adjust their leaves to better face the sun. This helps them absorb the most sunlight for photosynthesis. Some types of flowers can also open and close their petals or fold up their leaves in the evening in response to the environment around them.

So, scientists define a living thing as something that grows, moves, requires nutrients, and has respiration. Is there anything else we can use to identify a living thing?

Actually, there is. Living things also reproduce. People, birds, animals, and fish reproduce by having babies or laying eggs. Plants and trees reproduce through seeds. We see in Genesis 1:28 that this is part of God's design:

*God blessed them and said to them, "Be fruitful and increase in number; fill the earth and subdue it. Rule over the fish in the sea and the birds in the sky and over every living creature that moves on the ground."*

Name: ______________________

God designed living things to reproduce and increase in number. One thing we notice in living things is that they reproduce after their kinds — we'll be talking more about that next week!

We've learned five traits we can use to identify living things. Write those five traits below. You can look back in this lesson and in Day 2's lesson if you need to.

1. ______________________

2. ______________________

3. ______________________

4. ______________________

5. ______________________

It's been fun to learn about living things this week, and I'm excited to continue our exploration of biology next week! In the meantime, I have a question. We learned that scientists organize creation into living and nonliving things. We also learned that living things must breathe, grow, move, eat, and reproduce. Human beings are living things — but what makes us different from the plants, animals, birds, and fish that are also living things?

Excellent question, Ben! Though science can help us organize and define what a living thing is, it can't tell us why humans are different than other living things. To answer this question, we need to turn back to the Book of Genesis. Do you remember what we learned on Day 1 this week, Ben?

Yes — we studied what God created on each day.

Let's review what the Bible says about the living things God created. In Genesis 1:11, it says this about the third day of creation,

*Then God said, "Let the land produce vegetation: seed-bearing plants and trees on the land that bear fruit with seed in it, according to their various kinds." And it was so.*

On the fifth day of creation, God created the living things that dwell in the water and in the air. In other words, birds and marine life! In Genesis 1:20 it says,

*And God said, "Let the water teem with living creatures, and let birds fly above the earth across the vault of the sky."*

On the sixth day, we read that God created the living creatures that dwell on the land. Genesis 1:24 says,

*And God said, "Let the land produce living creatures according to their kinds: the livestock, the creatures that move along the ground, and the wild animals, each according to its kind." And it was so.*

But that wasn't all God created on the sixth day! Let's read Genesis 1:26–27 very carefully together. What do you notice that is different?

*Then God said, "Let us make mankind in our image, in our likeness, so that they may rule over the fish in the sea and the birds in the sky, over the livestock and all the wild animals, and over all the creatures that move along the ground." So God created mankind in his own image, in the image of God he created them; male and female he created them.*

Hmm, it says that God created mankind in His image — He didn't create plants or other living things in His image.

Exactly! We've talked before about what it means to be made in the image of God. In Genesis 2:7 we also read a few more details about the creation of mankind,

*Then the LORD God formed a man from the dust of the ground and breathed into his nostrils the breath of life, and the man became a living being.*

God only made mankind in His image and gave us the breath of life. This is what makes human beings different from all the other living things God created.

Look up Genesis 2:7 in your Bible. If you'd like, you can highlight this verse in your Bible. Memorize Genesis 2:7 with your teacher or with a sibling.

Did you have fun learning about what makes a living thing a living thing this week, friend?

I sure did — and now I'm ready to add a new page to my Science Notebook! God created so many different types of living things. I was thinking it would be fun to draw a picture of our favorite living things. It could be a person, animal, plant, bird, fish, lizard — anything that is classified as a living thing.

That sounds like fun! We can use our imaginations to help us draw our favorite living things. Or we can find a picture of our favorite living thing in a book or magazine.

Here is what our drawings look like. I drew a picture of a cat. Hannah loves flowers, so she drew a field of flowers, and Sam drew his favorite dinosaur. We can't wait to see what your favorite living thing is!

In your Notebook, write: Living things move, grow, breathe, reproduce, and need nutrients.

Then draw a picture of your favorite living thing.

Learning about living things this week also reminded us that God created human beings different from all the other living things. Copy Genesis 2:7 on the back of your Notebook page as a reminder.

*Then the Lord God formed a man from the dust of the ground and breathed into his nostrils the breath of life, and the man became a living being* (Genesis 2:7).

week 19

# Classification

Day 1

Hey there! I'm glad you're back for another adventure in biology with us. We had fun learning how we can identify living things last week. Living things all have several traits in common; do you remember any of them?

Ooh, I do! All living things move, grow, breathe, reproduce, and need nutrients.

Perfect! You may remember that we've talked before about how science is all about organization. Once we've identified a living thing, we can use science to help us organize it into a group of similar living things.

In the past, we've organized living things as human, animal, bird, fish, plant, or reptile. Is that what you mean?

That's one way we can do it — but we can organize living things in much more detail than that! **Classify** (said this way: klăs-ŭh-fī) means to organize or arrange things. The field of biology helps us to classify all living things by features or traits that they have in common.

**Taxonomy** (said this way: tăk-sŏn-ŭh-mē) is the field of science that deals with classifying living things — and it's part of biology! Through taxonomy, we classify living things by kingdom, phylum (said this way: fī-lŭhm), class, order, family, genus (said this way: jē-nŭhs), and species.

Whoa, that's a lot!

It sounds like a lot, but don't worry! We'll walk through it together. You can think of taxonomy's classification system like a funnel. A funnel is wide at the top but small at the bottom. In the same way, the classification system is broad at the top but specific at the end.

In other words, a kingdom is very broad, but the species is very specific. As we move through the classification system, we move from the broad kingdom and get more and more specific until we classify the species a living thing belongs to.

Would you like to classify a living thing together?

Sure! Let's start with our cat, Bell.

Great idea. So first, we'll need to determine what kingdom Bell belongs to. When we begin to classify something, we first need to ask ourselves, "Is this a plant or animal?" Our answer to this question tells us what classification kingdom the living thing belongs to. Bell is an animal, which means she is part of the animal kingdom. In taxonomy, the animal kingdom includes animals, birds, fish, reptiles, and humans.

But wait, humans aren't animals! God created human beings in His image, different from the animals.

Good catch! We'll talk about why humans are classified this way tomorrow. In the meantime, let's practice classifying some living things.

Cut out the living things on the next page then glue them in the plant or animal kingdom to classify them.

# Is This a Plant or Animal?

| Plant Kingdom | Animal Kingdom |
| --- | --- |
| | |

Blank for cutting purposes.

Welcome back, friend! We began learning about the field of taxonomy yesterday, and we talked about the animal kingdom. We learned that the animal kingdom includes animals, birds, fish, reptiles, and humans. Can you tell us why humans are included in the animal kingdom, Ben?

The answer lies in the two different worldviews we've been talking about. Biologists work to organize the living things we see on earth — but not all scientists begin with the Bible. In a secular, or evolutionary, worldview, a scientist would not believe that God is the Creator. Instead, they would believe that humans came from animals, slowly changing from animal to human over a long period of time through evolution.

When we begin with the Bible as the foundation of our worldview, we see that God created animals and humans differently. Only human beings were created in His image — and that separates us from the animals.

As we explore taxonomy together, it's important to recognize that God's creation is very organized, and we can organize living things by the similarities they have. We can also recognize that though taxonomy classifies humans in the animal kingdom, the Bible tells us that humans are made in God's image — to reflect His character and attributes — and that is what makes people very different from the animals.

Thanks for explaining that, Ben. Let's get back to classifying Bell now. We placed her in the animal kingdom yesterday. Now we need to learn what phylum she belongs to!

The phylum helps us to divide living things by certain traits they have in common. When we classify an animal into a phylum, we ask, "Does this animal have a backbone or not?"

Ah, the backbone, or spine, in a living thing is made of bones called vertebrae (said this way: vur-tŭh-brŭh). Living things that have a backbone are called **vertebrates** (said this way: vur-tŭh-brŭhts). If a living thing does not have a backbone, it is called an **invertebrate**. Our cat Bell does have a backbone, so she would be a vertebrate.

That means her kingdom is animal and her phylum is vertebrate.

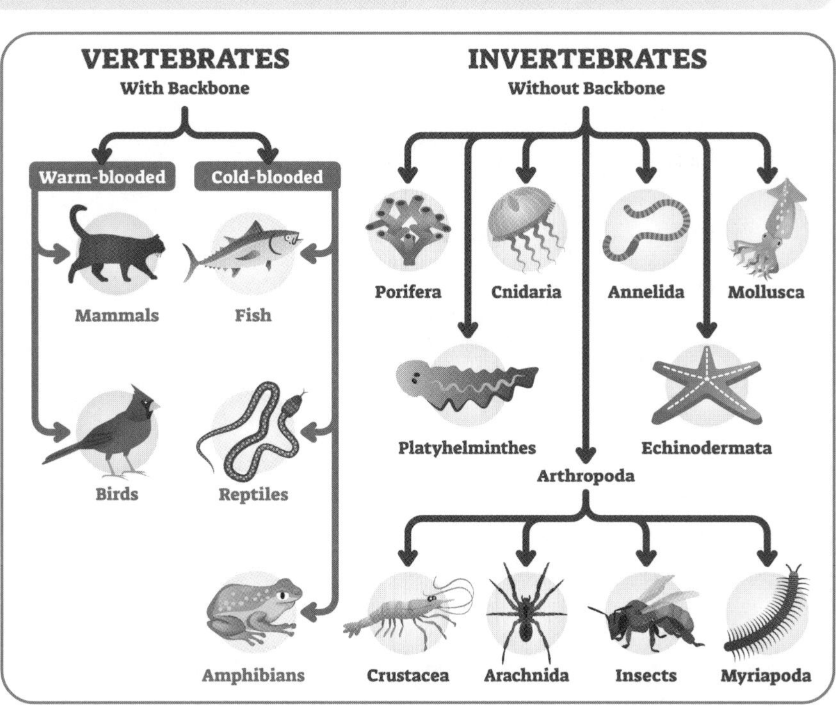

Name: ______________________________

Okay, we have the kingdom and phylum now. Class is next!

Ah, and we're out of time for today — but we'll be able to finish classifying Bell tomorrow. See you then, friend!

apply it

1. Do you remember Bell's kingdom and phylum? Write them below.

**Kingdom:** ______________________________

**Phylum:** ______________________________

2. Ask your teacher or a sibling to let you feel their backbone in the center of their back. What does the backbone feel like?

______________________________

______________________________

______________________________

______________________________

______________________________

______________________________

3. What separates humans and animals?

______________________________

______________________________

______________________________

______________________________

We're using taxonomy to classify our cat, Bell, this week. So far, we've learned that Bell's kingdom is animal and her phylum is vertebrate. Class is next! Just like phylum, class organizes living things by more specific traits that they have in common. A few classes in the animal kingdom are birds, reptiles, and mammals.

Mammals are living things that have hair on their bodies, feed their babies milk, and are vertebrates. Bell is definitely a mammal.

Right! Next, we need to classify Bell's order. The order divides living things even further by traits they have in common. Bell's order is carnivore; a carnivore eats meat. Next, we classify Bell's family. Dogs fit into the Canidae (said this way: kă-nŭh-dē) family, while cats fit into the Felidae (said this way: fē-lŭh-dē) family.

After family comes genus. Bell's genus is *Felis*. The genus *Felis* classifies cats that are on the smaller side. Finally, her species is *catus*. This species classifies domestic cats. Domestic is a word that refers to the home or household. Domestic cats typically live with people in or around their home.

Yay, we've classified Bell using taxonomy! Let's look at what we've written down in our Notebook, Ben.

Do you see how this system moves from a broad classification to a specific one? This classification also helps us gather information about Bell quickly. For example, by looking at her classification, we can see right away that she is a carnivorous mammal that has a backbone.

*Felis catus* seems like a funny way to say a cat that lives with people, though!

I'm glad you mentioned that. Do you remember when we talked about measurements and how science needs to be consistent, no matter where a scientist is? As scientists work to name and classify living things, they also need consistent names.

Ooh, that could be a problem since there are different languages all over the world. We call Bell a cat, but in the Spanish language, she would be called *gata*. In France, she would be called a *chatte*. The different languages scientists speak around the world could make classification really confusing!

Right. That is why scientists chose to use the Latin language to name and classify living things in science. Latin is a language that isn't commonly used anymore — but it is used in science to keep names consistent no matter where the scientist lives.

Name: ______________________

Find each word in the word search:

carnivore vertebrate invertebrate mammal domestic

Latin animal

| | | | | | | | | | | | |
|---|---|---|---|---|---|---|---|---|---|---|---|
| Z | V | E | R | T | E | B | R | A | T | E | M |
| C | A | C | A | R | N | I | V | O | R | E | A |
| A | N | I | M | A | L | Q | F | I | J | L | M |
| C | A | R | U | A | A | C | T | U | R | E | M |
| J | Z | I | T | R | E | N | T | O | R | Z | A |
| P | R | I | D | O | M | E | S | T | I | C | L |
| I | N | V | E | R | T | E | B | R | A | T | E |

Welcome back, friend! I've had fun exploring taxonomy with you this week.

I thought it was really interesting how we are able to classify living things. It reminds me of how organized God's creation is. The field of taxonomy also reminds me of what we read in Genesis 1 about the different kinds of living things God created.

Oh, I remember reading about kinds in Genesis. Let's review what Genesis 1:20–23 says,

*And God said, "Let the water teem with living creatures, and let birds fly above the earth across the vault of the sky." So God created the great creatures of the sea and every living thing with which the water teems and that moves about in it, according to their kinds, and every winged bird according to its kind. And God saw that it was good. God blessed them and said, "Be fruitful and increase in number and fill the water in the seas, and let the birds increase on the earth." And there was evening, and there was morning — the fifth day.*

We also see kinds mentioned when God gave instructions to Noah before the worldwide Flood in Genesis 6:19–20,

*You are to bring into the ark two of all living creatures, male and female, to keep them alive with you. Two of every kind of bird, of every kind of animal and of every kind of creature that moves along the ground will come to you to be kept alive.*

Those verses lead us to a question: what is a kind?

Great question, Ben! A **kind** is a group of living things that have similar traits and can reproduce. Hmm, let's think about horses.

I'm thinking of the horse farm we visited last week. They had all different kinds of horses there. Each horse looked unique — but you could still tell they were horses. And the foals, or baby horses, were so cute!

Right, we observed the horse kind that day. Another kind of animal is the dog kind. The dog kind includes domestic dogs, wolves, coyotes, and even foxes. Each animal within a kind may look different — this is called variation.

We know a lot of different dogs. Some are big; others are very small. Our neighbor's dog has spots, but Uncle Gus' dog has stripes.

This is variation within the dog kind. But, just as the Bible says, a kind reproduces after its own kind. Even though we may see variation like different colors or sizes within a kind, a dog always stays a dog, a cat always stays a cat, and a duck always stays a duck.

Look up Genesis 1:21 in your Bible. If you'd like, you can highlight this verse in your Bible. Memorize Genesis 1:21 with your teacher or with a sibling.

It's the best day of the week — we get to add a new page to our Science Notebook today!

Wasn't it fun to learn about how organized God's creation is and how we can organize living things? I enjoyed learning about taxonomy with you.

Me too! Since we classified Bell, let's draw a picture of her in our Science Notebook. We can write down how she is classified by taxonomy.

Ooh, I like it! Friend, if you have a cat of your own, you can draw a picture of them. Your pet cat is classified just the same as our cat, Bell.

Here is how each of our Notebooks turned out! Doesn't Sam's cat look happy? And I love the unique color Ben gave his cat. Have fun creating your Notebook, friend!

In your Notebook, write:

**Cat**

Kingdom: Animal
Phylum: Vertebrate
Class: Mammal
Order: Carnivore
Family: Felidae
Genus: *Felis*
Species: *Felis catus*

Then draw a picture of a cat.

Learning about taxonomy this week also reminded us that God created many plant and animal kinds. Just as the Bible says, each kind of animal reproduces after its own kind. A dog is always a dog, for example. Copy Genesis 1:21 on the back of your Notebook page as a reminder.

*So God created the great creatures of the sea and every living thing with which the water teems and that moves about in it, according to their kinds, and every winged bird according to its kind. And God saw that it was good* (Genesis 1:21).

week 20

# Day 1 Marine Biology

Hi there! Hannah and I are getting ready to start our science adventure for today, and we're glad you've come back to join us.

Uh, Ben — we have a bit of a problem here.

Nah, we already finished all of our math problems for the day. We don't have any problems left to do.

No, not a math problem. We have an adventure problem! We're exploring biology together now — but there is so much we could explore. Just think about it — we could learn about the human body, animals, flowers, trees, plants, ocean life, cells, ecosystems, and more. Our problem is, what do we explore on our adventure now?

Oh, I see what you mean. We can't possibly explore all there is to know about biology during the rest of our adventure together this year. I think we should pick a topic and then dive deep into exploring it. Hey, that gives me an idea! I've always wanted to learn more about the ocean and the life it contains. What do you think, Hannah?

Ah, we could focus on exploring the ocean through marine biology together!

This is going to be so exciting! I have the best ideas, if I do say so myself.

Oh, Ben! Let's begin exploring marine biology together — are you ready to join us, friend? I guess we should start by learning what marine biology is.

Well, the word marine refers to the sea or ocean. So, marine biology is the field of biology that explores the ocean. What kind of scientist explores the ocean, Hannah?

**Weekly materials list**

- [x] Kitchen scale that measures in grams
- [ ] Plate
- [ ] Water
- [ ] Salt
- [ ] Spoon
- [ ] Pitcher or container

Name:

A marine biologist! A marine biologist is a scientist who studies the ocean and the life we find in it. These scientists work to classify marine life, understand their behavior, monitor and explore the ocean, and sometimes even discover new kinds of living things! A marine biologist must understand portions of chemistry, biology, and even the geology of the ocean in order to do their work.

Wow, it sounds like there is a lot we can learn about together in the field of marine biology. I'm excited to learn more about this field of science as we continue our adventures together.

1. What does a marine biologist study?

2. What do you think it would be like to be a marine biologist?

3. What is something you would like to learn about the ocean or marine life?

We're back together, and it's time to begin our exploration of marine biology! I hope you remember some of what we learned in chemistry because we're going to be talking about it again today.

I guess I better dig out my periodic table of elements, then. Let's start with the ocean. What is the ocean?

**materials needed**

- Kitchen scale that measures in grams
- Pitcher or container
- Spoon
- Water
- Salt

An ocean is a very large expanse of water. Though all of the oceans on earth are really connected to each other, we divide them into five oceans.

Hmm, kind of like the way a large area of land, like the United States of America, is divided into different states. Let me see if I can remember the five oceans . . . there is the Atlantic Ocean, Arctic Ocean, Indian Ocean, Pacific Ocean, and the Southern Ocean.

Correct! Did you know that the oceans actually cover most of the earth? It's true! Over 70% of the earth is covered by ocean.

Wow — so there is more ocean than land on the earth? That means there is definitely a lot we can learn about and discover in the ocean.

Hannah, do you remember the trip we took to the ocean a few years ago? We had so much fun playing in the sand and in the water. We even saw a dolphin swimming through the water from our hotel room window!

We did have a lot of fun — until I got a big mouthful of ocean water. Yuck!

Hmm, that gives me a question. What is the ocean made of? What makes it different from the water we drink?

Good question. First, the ocean is made from water. Lots and lots of water! Do you remember the chemical formula for water, Ben?

Sure do! It is $H_2O$, which means a water molecule is made from two hydrogen atoms and one oxygen atom bonded together. But I drink good old $H_2O$ all the time, and it doesn't taste bad like the ocean water did. So, what else is in ocean water?

Ocean water contains salt. Salt can also be called sodium chloride. A molecule of salt is formed by one atom of sodium and one atom of chlorine bonded together. The chemical formula for salt is NaCl.

Name: ______

Ah, that explains the taste. Is there a lot of salt in the ocean?

Ocean water is about 3.5% salt. That gives me an idea — let's make our own sample of ocean water.

## Activity directions:

1. Place the kitchen scale on the table then set your container on top of it.
2. Turn on the scale — make sure it reads 0.00. If it does not, press the "Tare" or "Zero" button. Be sure the scale is set to measure in grams.
3. Pour water into the pitcher until the scale reads 965 grams of water.
4. Now, add salt to the pitcher until the scale reads 1,000 grams.
5. Mix the water and salt together until the salt has fully dissolved. You've created a sample of ocean water! **Save some of your ocean water in a container for tomorrow's lesson, then fill in the worksheet with your observations.**

ASK PARENT FOR HELP

**Note**

It's okay to take a small taste of your ocean water — but do not drink more than a small taste.

apply it

1. What does your ocean water smell like?

______

2. Use a small spoon to take a small taste of your ocean water. What does it taste like?

______

3. Lakes, rivers, ponds, and streams are what we call fresh water because there is very little salt in them. Compare your ocean water to a glass of fresh water you would drink. Do they look different?

______

4. How does the water you drink taste different from the ocean water?

______

______

______

What did you think of our ocean water activity yesterday, Hannah?

Yuck! It definitely reminded me of that mouthful of ocean water I got at the beach.

The water in the ocean is called salt water because of the salt it contains. It sure doesn't taste good, but why else wouldn't we want to drink salt water, Hannah?

**materials needed**

- Plate
- Ocean water from Day 2

Salt is an important nutrient to our bodies — but too much salt is not a good thing. If we consume too much salt, it can cause our body to dehydrate (said this way: dē-hī-drāt), which means to lose too much water.

Ah, so drinking salt water from the ocean wouldn't give us the water our body needs. Instead, it would leave us thirstier and more dehydrated than before!

Exactly right. That is why we only drink fresh water to stay hydrated. When we are hydrated, it means that our body has the water it needs to survive.

That is good to know. So, is water and salt all that the ocean is made of?

Well, there are small amounts of other elements like carbon, potassium, and calcium in the ocean. But the majority of the ocean is made from simple water and salt.

While we were eating dinner last night, I noticed that we have sea salt at the table. Did that salt come from the ocean? If it did, how can salt be gathered from the ocean since it is dissolved in the water?

Ah, great questions, Ben. Yes, sea salt is gathered from the ocean or from lakes with salt water. Do you remember when we talked about mixtures and solutions a few weeks ago? Salt water is a solution. Remember, a solution is formed when a substance is dissolved into a different substance.

So in this case, water is the solvent that dissolves the salt — and the salt is the solute!

You got it! With salt water, we can separate the solution back into water and salt through evaporation. To collect sea salt, salt water can be gathered into an area. Then it is left alone while the sun evaporates the water. Once the water has evaporated, just the salt is left to be gathered and sold to our grocery store.

Hey, let's evaporate some of the ocean water from our last activity to separate the salt and water!

## Activity directions:

ASK PARENT FOR HELP

1. Write your name and the date on your lab report. Next, write the question, "How can we separate a solution of salt and water?"
2. Write your hypothesis in the Hypothesis section of your lab report. Ben and Hannah's hypothesis was, "Evaporation can be used to separate the water and the salt."
3. Decide where you want to set your plate. If it is a warm, sunny day, you can do this activity outside. Otherwise, you can set your plate on the kitchen counter.
4. Pour some of the ocean water onto the plate. The water should just cover the bottom of the plate — but don't cover it too deep or it will take a long time for all of the water to evaporate.
5. Now it is time to wait for the water to evaporate — it may take a day or more for all of the water to evaporate from the plate. Write anything you observe, like how long it takes the water to evaporate, in the "Things I observed" section of your report.
6. Once the water has fully evaporated, examine the plate. What is left behind? Finish your lab report and record the results of your experiment.

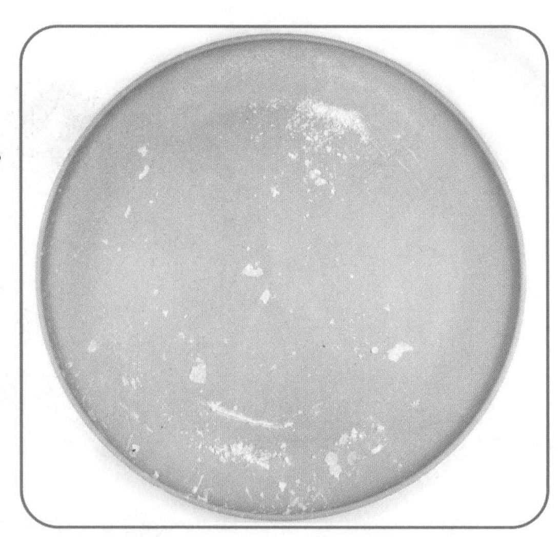

Name ______________________ Date ______________

# Lab Report

## Question

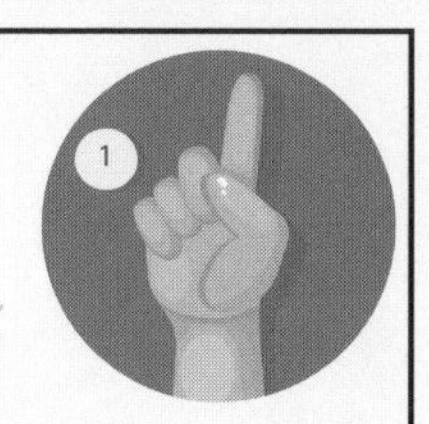


______________________

______________________

______________________

______________________

## Hypothesis

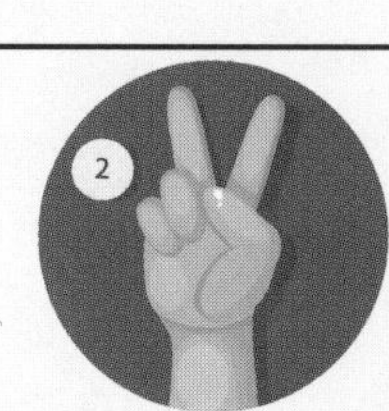


______________________

______________________

______________________

______________________

## Things I observed:

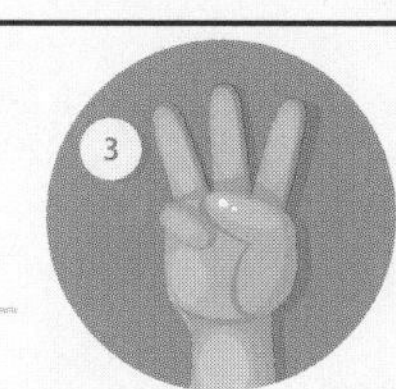


______________________

______________________

______________________

______________________

## Results

### What happened in the experiment?

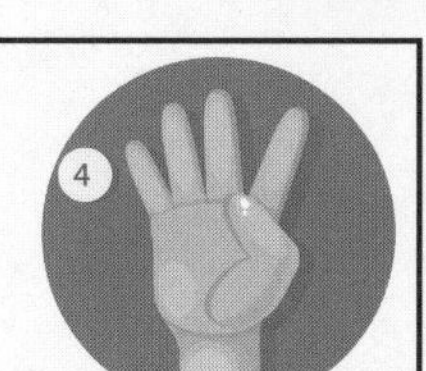


______________________

______________________

______________________

______________________

______________________

______________________

### Was my hypothesis correct?

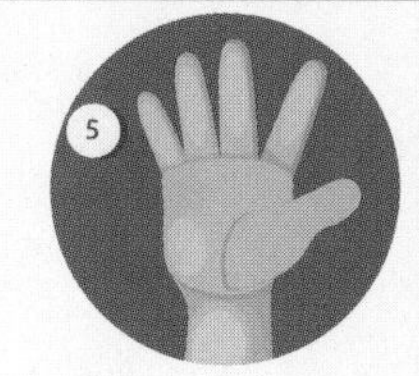


______________________

______________________

# Additional Lab Notes

# Day

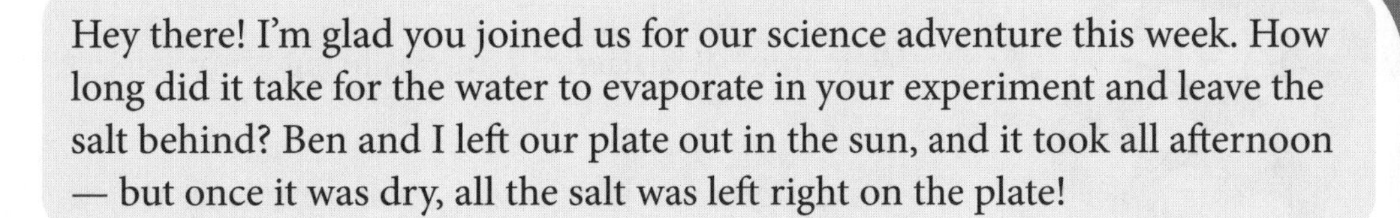

Hey there! I'm glad you joined us for our science adventure this week. How long did it take for the water to evaporate in your experiment and leave the salt behind? Ben and I left our plate out in the sun, and it took all afternoon — but once it was dry, all the salt was left right on the plate!

I enjoyed learning more about ocean water this week. I've been thinking about how much water is on the earth. I remember standing on the beach and looking out to the ocean — it looked endless to me! Sometimes the waves were also so big and powerful, they felt a little scary at times.

That reminds me of Matthew 8:23–27. Jesus and his disciples were in a boat on a lake when a big storm passed through. Have you ever had a big storm suddenly pass through your area, friend? Well, this storm came suddenly, and soon the waves were crashing over the boat!

I'd be terrified!

And that is exactly what the disciples were — let's read what Jesus' disciple Matthew wrote about that day.

*Then he [Jesus] got into the boat and his disciples followed him. Suddenly a furious storm came up on the lake, so that the waves swept over the boat. But Jesus was sleeping. The disciples went and woke him, saying, "Lord, save us! We're going to drown!" He replied, "You of little faith, why are you so afraid?" Then he got up and rebuked the winds and the waves, and it was completely calm. The men were amazed and asked, "What kind of man is this? Even the winds and the waves obey him!""*

Whoa — it must have been amazing to see the waves crashing all around one minute and then completely calm the next!

Jesus had control, or authority, over the wind and the waves. He has authority over everything! When we go to the ocean or even just the pond in our neighborhood, the water reminds me that Jesus has authority over all things. No matter what is happening around me, I don't have to be afraid because He is with me.

What about you, friend? What is something that feels scary to you? Talk with your family about ways you can learn to trust God no matter what is happening around you.

Look up Matthew 8:26–27 in your Bible. If you'd like, you can highlight these verses in your Bible. Memorize Matthew 8:26–27 with your teacher or with a sibling.

Welcome back, friend! Did you bring your Science Notebook with you?

I did! I enjoyed learning about ocean water this week. Can we draw a picture of the ocean in our Notebooks this week?

I think that is a great idea! I have a picture we can use to give us ideas for our own drawings right here. If you'd like, you can also draw a boat in the water to remind you of the passage we read from Matthew 8.

These are our drawings. Sam and Hannah decided to add the reflection of the sunset on their water. We can't wait to see your drawing!

In your Notebook, write:

The ocean covers most of the earth.

Then draw a picture of the ocean.

Learning about ocean water this week also reminded us that Jesus has full authority over all things. Copy Matthew 8:26–27 on the back of your Notebook page as a reminder.

*He replied, "You of little faith, why are you so afraid?" Then he got up and rebuked the winds and the waves, and it was completely calm. The men were amazed and asked, "What kind of man is this? Even the winds and the waves obey him!"* (Matthew 8:26–27).

week 21

# Day 1 Ocean Movement

Oh good, you're here! We're exploring the ocean together, and I have an idea for what we can talk about this week.

Oh? What would that be, Ben?

Well, I'm still thinking about that trip we took to the ocean a few years ago. Do you remember the sandcastle we created together with Mom and Dad?

I do — we made such a neat castle! We even made a flag to sit on top of it. After we finished creating the castle, it was time to go eat. I didn't think we were gone for that long, but by the time we got back to the beach, the ocean had risen above our sandcastle! Our beautiful creation was under water, and the waves had washed our flag to the sandy shore.

- [x] Paint tray
- [ ] Water
- [ ] Small floating object
- [ ] Spoon
- [ ] Glass or plastic bottle with screw-on lid
- [ ] Blue food coloring
- [ ] Coconut or vegetable oil
- [ ] Funnel
- [ ] Plastic tablecloth

Weekly materials list

We were so disappointed — until Mom and Dad explained tides and waves to us. Then everything made sense. I was thinking we could explore tides and waves together with our friend this week!

Great idea, Ben. Like you said before, you're full of great ideas! Have you ever been to the ocean, friend? If you have, you may have noticed that the ocean doesn't sit still like a puddle of water. It is full of movement!

When you stand on the shore of the ocean, or even a lake or pond on a breezy day, you may notice the movement of waves through the water. Sometimes, the waves may look like just a small ripple on the surface of the water. Other times, the waves may swell much bigger — especially if it is stormy!

**Name:** ____________________

Waves are usually created by wind blowing over the surface of the water, though there can be other causes, such as a volcanic eruption or an earthquake. Volcanoes and earthquakes can cause large, very destructive waves.

Hey Ben, I have a puzzle for you! Finish this sentence: Waves are the movement of. . . .

Um, water?

Good guess! But waves are actually the movement of energy across the surface of the water. This movement of energy causes the water to swell and fall as the wave of energy passes through. I have an activity to help us see this in action.

## Activity directions:

1. Pour water into the paint tray — the deep end should have an inch or two of water in it.
2. Gently tap the deep end of the paint tray. After you tap it, you should see waves of energy ripple through the water to the shallow end of the tray. If you don't see waves, tap a little harder until you do see waves ripple through the water. Answer question 1 on your worksheet below.

3. Now, place your floating object gently into the deep side of the paint tray.
4. Once the water has settled again, tap the deep end of the tray to create waves. Observe the floating object. Does it tend to stay in the same place, or is it moving along the waves? Answer question 2 on your worksheet.
5. Now, use the spoon to create waves in the water. You can push or pull the water in the tray. What happens to the floating object now? Answer question 3 on your worksheet.

1. Did you see the wave of energy travel through the water after tapping the paint tray? Yes / No
2. Did your floating object tend to stay in the same place as the wave of energy passed underneath, or did it travel along the path of the wave?

____________________

3. What happened when you used the spoon to move the water — did your floating object move along with the water or stay in the same place?

____________________

Our activity with the paint tray sure was fun! I noticed that when we created waves by tapping on the tray, our floating object tended to ride over the wave as the energy passed by underneath. But once we moved the water with the spoon, then the floating object moved along with the water. I'm still a little confused, though — can you explain how what we saw in the tray is similar to ocean waves?

**materials needed**

- Glass or plastic bottle with screw-on lid
- Water
- Blue food coloring
- Coconut or vegetable oil
- Funnel
- Plastic tablecloth

When we tapped on the tray, it created a wave of energy that traveled over the surface of the water — just like the wind creates the wave of energy that travels over the surface of the ocean. While the wave of energy causes the water to rise and fall, it doesn't move the water along its path. That's why the floating object stayed in about the same place — the wave of energy passed underneath it and left the water in place.

Hmm, so it's like if I were to hold my bedsheet by one side and flap it. This would cause a wave of energy to travel along the sheet, making the sheet rise and fall. But the sheet would still stay in my hands and would settle in about the same place once the wave of energy passed through.

Right! But when we used the spoon to push or pull the water, we were causing the water itself to move. Because the water was also traveling along the path of the wave, this caused the floating object to move along with the water.

That makes sense! So, if waves are energy passing over the water, what causes waves to pull the water higher as they get closer to the shore?

As the wave travels closer and closer to the shore, the water becomes shallower and shallower. Shallow means the water isn't as deep. The closer you are to the shore, the shallower the water is because of the land rising out of the water.

As the water becomes shallower, the land underneath the water begins to create friction against the movement of the wave. **Friction** (said this way: frĭk-shŭhn) is a type of force that pushes against movement. The friction that land creates begins to slow the bottom of the wave down — but the top of the wave is still traveling quickly. This causes the water to pull up into a taller wave. Once the friction from the land becomes too great for the wave, the wave collapses or crashes against the shore.

I'm looking forward to the next time we get to watch some waves roll into the shore. In the meantime, let's create some waves in a bottle today.

## Activity directions:

1. Spread out a tablecloth to protect the surface you are working on.
2. Use the funnel to pour water into the bottle. Fill the bottle up about halfway.
3. Add 2–3 drops of blue food coloring to the water to create the ocean. Screw the lid back onto the bottle and gently swirl the bottle to mix the water and food coloring. If you would like your ocean to be darker, you can add another drop or two of food coloring.
4. Remove the lid. Use the funnel to fill the rest of the bottle with oil. Be careful not to overfill it!
5. Screw the lid onto the bottle tightly. Ask an adult to check to make sure it is secure.
6. Hold the bottle sideways by each end and tilt it from side to side. Observe the movement of the waves in your bottle. Have fun!

Now that we've learned about waves, it's time to talk about the tide. The tide is part of the ocean's movement, and it affects how deep the water is on the shore. Let's imagine that we are looking out at the ocean shore as I explain tides. Is your imagination ready?

When we first look at the shoreline, we notice that there isn't a lot of sand to play on. The water is quite high and has covered most of the beach. This would be called the high tide. But when we return a while later in the day, the beach suddenly looks much bigger! The water is now farther back, and there is more sand exposed for us to play on. This would be called the low tide.

Interesting! So, the tide changes the level of the water along the shoreline each day. What causes the level of water to change between high tide and low tide, Ben?

I'm glad you asked! The moon travels in an orbit around the earth due to the earth's gravity pulling against it. But did you know that the moon also has gravity that pulls against the earth? It's the moon's gravity pulling against the earth that causes the tides we see.

As the moon travels along its orbit, its gravity pulls against the earth. Though you and I can't feel this gravitational pull, it does affect the earth and the ocean. As the moon's gravity pulls against the side of the earth that is facing it, the gravity causes the water to gather up or bulge on this side of the earth. This is what causes high tide.

But that's not all! The gravitational pull of the moon also affects the earth and causes it to shift slightly too. This causes the water to gather up or bulge on the opposite side of the earth too. This means that in most areas of the earth, there will be two high tides and two low tides each day as the moon orbits around the rotating earth.

When the water level is low during the low tide, it exposes more of the shoreline. In some places, the low tide may only show us a few more feet of the shore. In other places, a low tide can expose many feet of shore and reveal new, interesting features that were covered in water during the high tide.

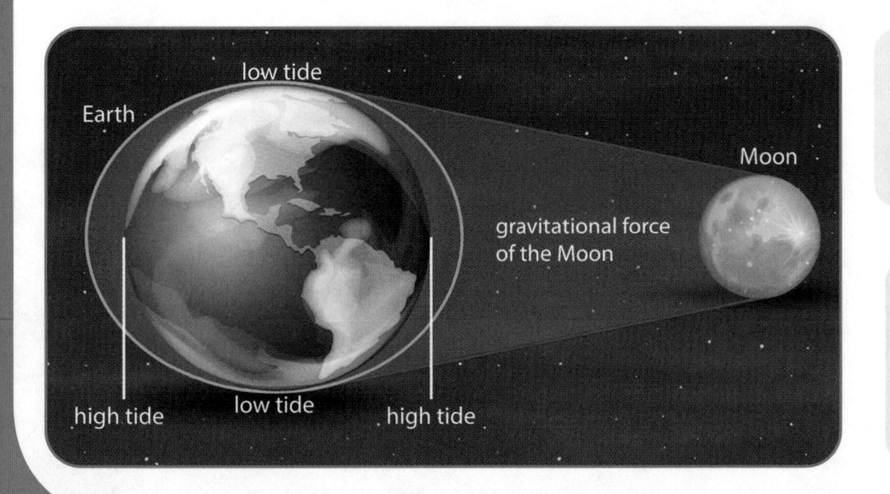


Wow, thanks for explaining that to us, Ben! Is the moon the only thing that affects the tides on earth?

No, the sun's gravity also affects the tides. But because the moon is closer to the earth, it has the biggest impact on tides.

Name: ______________________________

1. What causes the tides on earth?

______________________________

2. Look at the picture below. Use what you learned in the lesson to determine which areas of the earth would have high tide and which areas would have low tide. Label the high tides and low tides.

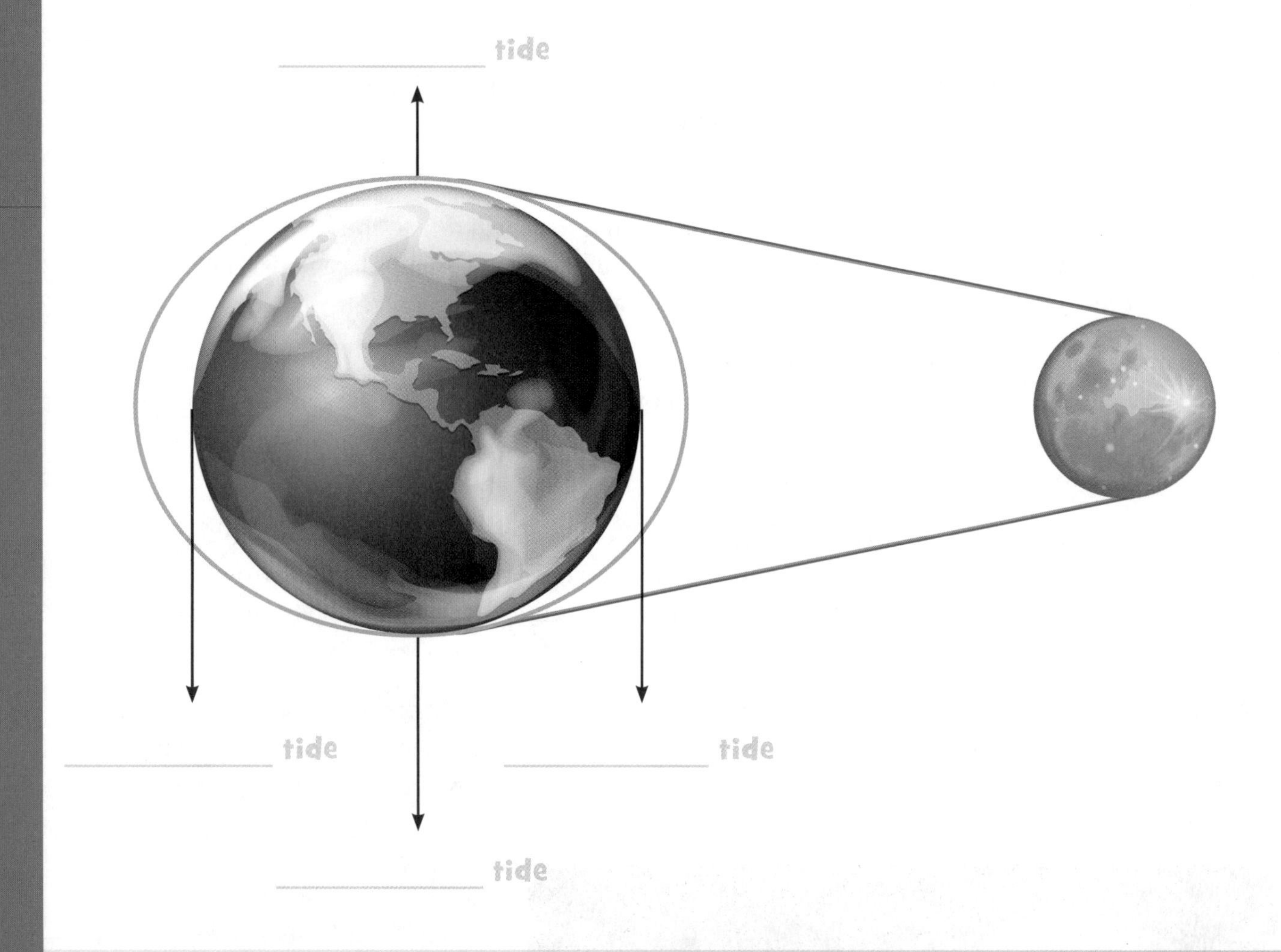

Hello again, friend! We've been talking about waves this week, and we even got to learn a little bit about tides. I'm looking forward to learning more about tides next week! But for now, learning about waves reminded me of Jonah's story in the Bible.

Oh, I remember his story! Jonah was a prophet. God gave prophets an important job in the Bible. They were to deliver God's messages or warnings to the people. God told Jonah that he was to go to the city of Nineveh and tell the people living there that God saw their great wickedness and that He was going to destroy the city.

But for Jonah, there was a problem! He didn't want to go to Nineveh, and he didn't want to warn the people. So rather than listen to God, Jonah decided to travel in the opposite direction! He thought that he could run away from God and the job he had been given.

But there was another problem — we can't hide from God! As it says in Psalm 139:7–8,

*Where can I go from your Spirit? Where can I flee from your presence? If I go up to the heavens, you are there; if I make my bed in the depths, you are there.*

Jonah got onto a boat as he ran away from God. Then God sent a huge storm, and the sailors were scared the boat was going to break apart as the waves crashed around them! Jonah told the sailors that he was running from God's directions and that the storm was a consequence of his decisions. Jonah told the sailors to throw him into the sea and that the storm would calm.

Though they didn't want to throw Jonah overboard, the storm continued to grow worse and worse until finally, they decided to do as Jonah had said. Then the sea grew calm, and God sent a huge fish to swallow Jonah. God had mercy on Jonah and kept him alive inside that great fish! There, Jonah prayed to God,

*"In my distress I called to the Lord, and he answered me. From deep in the realm of the dead I called for help, and you listened to my cry"* (Jonah 2:2).

God showed mercy to Jonah, and the fish returned him to dry land. This time, Jonah obeyed God and went to Nineveh — but he still wasn't happy about it. Because Jonah obeyed God and warned the people of Nineveh, though, the people asked God for forgiveness. They repented for all of their wickedness — and God had mercy on them too.

The story of Jonah and the Ninevites reminds us that God shows us mercy when we repent. He forgives us no matter what mistakes we have made. It also reminds us to show mercy to others and to care about them just as God does.

Hey, let's read Jonah's full story tonight!

Read the Book of Jonah together with your family. What lessons can you learn from Jonah's story? Look up Psalm 139:7–8 in your Bible. If you'd like, you can highlight these verses. Memorize Psalm 139:7–8 with your teacher or with a sibling.

It's Science Notebook day, woohoo! We've been learning about ocean waves and tides this week. It has been a lot of fun. Ben has enjoyed creating waves in our ocean bottle; how about you, friend?

I sure have! I also have an idea for our Science Notebook this week. Since the moon has the greatest impact on tides on the earth, let's draw a picture of the moon in our Notebook this week.

That sounds great! Here is an image we can use for an example. Let's get started.

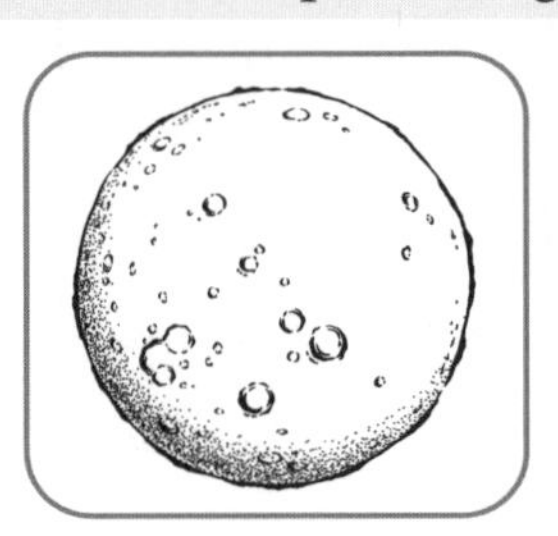

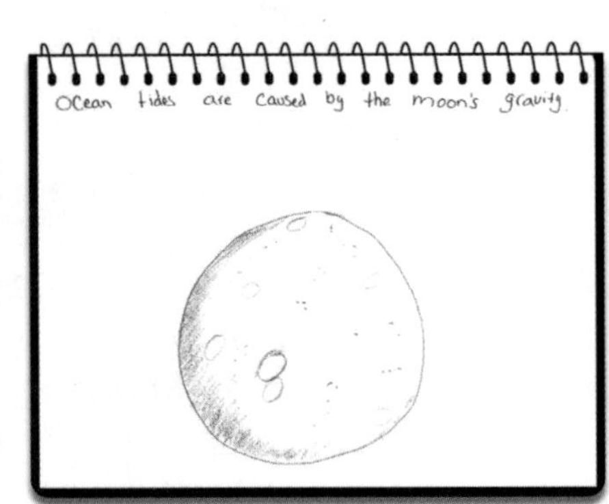

In your Notebook, write:

Ocean tides are caused by the moon's gravity.

Then draw a picture of the moon.

Learning about waves this week reminded us of Jonah's story and that we cannot hide from God. But no matter what mistakes we've made, God has mercy on us and forgives us when we repent. Copy Psalm 139:7–8 on the back of your Notebook page as a reminder.

*Where can I go from your Spirit? Where can I flee from your presence? If I go up to the heavens, you are there; if I make my bed in the depths, you are there* (Psalm 139:7–8).

week 22

# Tide Pools

# 1

Hey there, friend! We're going to continue exploring tides this week. I have a feeling we're going to discover some interesting things together.

Hmm, we'll see! But first, I've been reading more about tides and gravity since our last lesson. Do you know the name of the scientist who discovered and explained gravity? His name was Sir Isaac Newton.

Sir Isaac Newton was born in England in 1643, and God gave him a brilliant mind for math and science. In fact, Newton contributed so much to the fields of math and science that he is one of the most famous scientists in history! Sir Isaac Newton's work also involved explaining how tides are affected by the gravity of the moon and the sun. Through his work, he gave the world a greater understanding of science and of God.

Wait — he was a Christian?

He was indeed! The foundation of his worldview was the Bible. He believed that God created all that we see, just as the Bible describes in Genesis. Newton wrote on many different topics, including astronomy, gravity, mathematics, and the Bible. He said,

"I have a fundamental belief in the Bible as the Word of God, written by men who were inspired. I study the Bible daily."*

Wow, thanks for sharing that with us, Hannah! It is really interesting to learn a little more about scientists who held a biblical worldview.

I think so too. Now let's continue learning about tides together! Did you know that tides are important to the ocean? As water rushes into the shore and then back out during the tidal cycle, it helps to stir up the ocean's nutrients. The cycle of the tides also forms an interesting ecosystem that we can explore.

Hannah and I learned about ecosystems together in *Adventures on Planet Earth*! Let's review what an ecosystem is. An **ecosystem** (said this way: ē-cō-sĭs-tŭm) is a community of living and nonliving things that are together in one place. We can find an ecosystem in a pond, in the desert, or in the forest.

- [x] Air dry modeling clay
- [ ] Toothpicks
- [ ] Acrylic paint
- [ ] Paintbrush
- [ ] Paper plate
- [ ] Plastic tablecloth

**Weekly materials list**

*Source: J.H. Tiner, *Isaac Newton — Inventor, Scientist, and Teacher* (Fenton, MI: Mott Media, 1975).

Name: ____________________

Thanks, Ben! The cycle of high tide and low tide often creates an ecosystem in the intertidal zone. The **intertidal zone** (said this way: ĭn-ter-tī-dåhl zōne) is the area that is visible during low tide but under water during high tide. We'll begin exploring the intertidal zone tomorrow.

apply it

1. Copy Sir Isaac Newton's quote: "I have a fundamental belief in the Bible as the Word of God, written by men who were inspired. I study the Bible daily."

2. What is the area called that is visible during low tide but under water during high tide?

3. What is an ecosystem?

Welcome back, friend! We're going to explore the intertidal zone together today. But first, Ben, can you remind us how the tides are created?

Sure thing! The moon's gravitational pull causes water to gather or bulge on the side of the earth that is closest to the moon, as well as the opposite side. This creates the high tide on those two sides of the earth. When the tide is high, it covers more of the shoreline.

But what about the other two sides of the earth? The sides of the earth that are not in line with the moon will experience a low tide. During low tide, the water has receded, or moved away, from the shoreline. This exposes parts of the shore that we couldn't see before.

Remember, the parts of the shore that are exposed during low tide are called the intertidal zone. Sometimes, there are holes or piles of rocks in the intertidal zone that trap water as the tide recedes during low tide. This creates little puddles or ponds of water on the shore — we call them tide pools. Tide pools are often full of living things!

Hmm, when we were at the aquarium recently, there was a tide pool exhibit. They even let us touch the creatures in it!

Yes, an aquarium tide pool is similar to the tide pools we find in the intertidal zone. In an aquarium, the tide pool stays the same — but in nature, tide pools are a place of change.

During high tide, they are covered in water, but during low tide, there may be very little water. The sun may also evaporate the water in the pool and dry it out! Sometimes, tide pools are exposed to strong wind, rain, waves, and sunlight. The temperature can also change quite quickly as the ocean water rushes in and away. Tide pools are always changing.

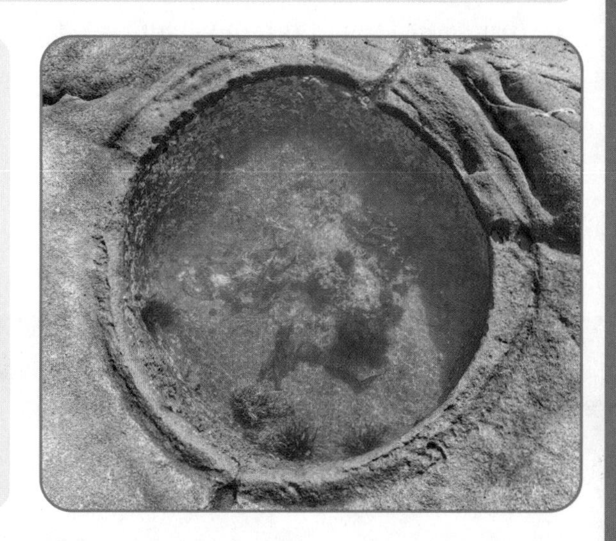

It sounds like a tide pool would be a tough place to survive. The creatures that live in this ecosystem must be able to adapt. **Adapt** (said this way: ŭh-dăpt) means to be able to adjust or change for certain conditions or a particular place.

For sure! A few of the living things that can be found in tide pools are sea stars, mollusks, crabs, algae, and more. We're going to learn more about these living things in our next few adventures together.

Name:

1. Have you ever been to an aquarium with a tide pool?

2. If so, were you able to touch any of the living things inside?

3. What does adapt mean?

4. Why must living things be able to adapt in a tide pool?

I'm so glad you're here now! Today, we're going to learn about one of my favorite living things that can be found in tide pools: sea stars! Sea stars can also be called starfish — but did you know they aren't actually fish at all?

Well, now I'm curious, what makes a fish a fish?

**materials needed**

- Air dry modeling clay
- Toothpicks
- Acrylic paint
- Paintbrush
- Paper plate
- Plastic tablecloth

Remember, we group living things through taxonomy by traits that they have in common. A few things that a living thing must have in order to be called a fish are that it must be a vertebrate, have fins, and use gills to breathe. Sea stars are invertebrates, which means they do not have a spine.

Interesting! What other creatures would sea stars be grouped with, then?

In taxonomy, sea stars belong in the animal kingdom, and their phylum is Echinodermata (said this way: ē-kī-nŭh-dur-mŭh-tŭh). Echinoderms include sea stars, sea urchins, and sea cucumbers. We'll talk more about those creatures next week!

There are just about 2,000 different species of sea stars, and they can be found in oceans all around the world. Sea stars come in many different colors and patterns, and some can be quite beautiful.

What kind of habitat do they live in?

Since they live all over the world, sea stars can be found in different marine habitats. A **habitat** (said this way: hăb-ĭ-tăt) is the natural environment a plant or animal lives in. A sea star's habitat includes tidal pools, rocky shore areas, and the coral reef. We'll be exploring the coral reef together soon!

Most species of sea stars have five arms, but some species can have 10, 20, 30, or even more arms! On the bottom of sea stars' arms are small, spiny feet that they use to move around or attach themselves securely to their surroundings. One interesting ability that God gave to sea stars is that if they lose one of their arms to an injury or to a predator, they can grow a new one. Isn't that neat?

Sea stars can often be found in tide pools. They are carnivores, which means they are predators that eat meat. A sea star typically eats oysters, mussels, clams, and snails. Some species of sea stars are keystone species because they help keep their ecosystem balanced.

A **keystone species** (said this way: kēē-stōne spē-shēz) is a type of animal that the ecosystem depends upon. If a keystone species disappears, the ecosystem changes in ways we'll notice over time.

Thanks, Ben. That's all the time we have for our adventure today — but we'll continue exploring the living things in tide pools next week.

## Activity directions:

1. Spread out a plastic tablecloth to protect the surface you're working on.
2. Roll out a small ball, about 1 inch, from the clay then flatten it to form the body of the sea star.
3. Roll out five arms from the clay.

4. Press the arms into the bottom of the round body. Carefully flip the sea star over.

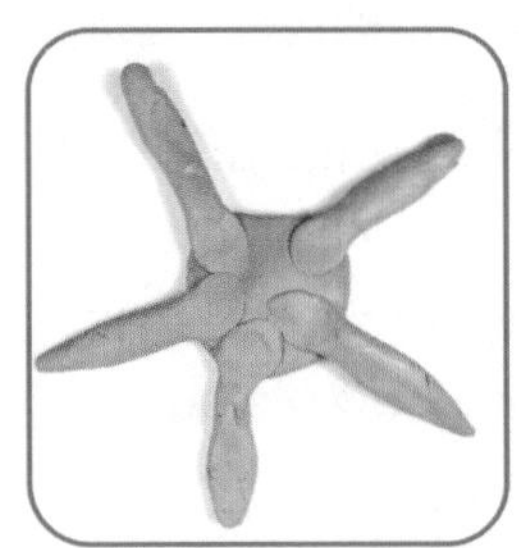

5. If you'd like, you can use the toothpick to draw lines or poke a design on the top of your sea star.
6. Carefully set the sea star on a paper plate and allow it to dry. Depending on how thick the clay is, it may take 1–3 days to dry.

7. Once it is dry, spread out the plastic tablecloth to protect the table. Paint your sea star.
8. Show someone your model sea star and share what you learned about them. Then place your sea star in a safe place and save it for next week's project.

It was fun to learn about tide pools this week! I have a feeling we have so much more to learn about them as we continue our science adventure next week.

We sure do! In the meantime, though, our discussion about sea stars reminded me of something. The word regenerate means to regrow, renew, or to bring about something new. God gave sea stars the ability to regenerate an arm if they lose one. This reminded me of how God renews and regenerates us.

Before we decide to follow Jesus, the Bible says in Ephesians 2:2 that we are dead in our sins. We have no life in ourselves, and there is nothing we can do about it. But when we repent of our sin and place our trust in Jesus, He gives us new life and cleanses us from the stain of sin. Ephesians 2:4–5 explains it this way:

*But because of his great love for us, God, who is rich in mercy, made us alive with Christ even when we were dead in transgressions — it is by grace you have been saved.*

Grace is the free gift from God. We don't deserve His grace, and there is nothing we can ever do to earn it — He gives it freely to us. Through God's love, mercy, and grace, He takes away that old, dead sin nature and gives us a new life. In 2 Corinthians 5:17 it says,

*Therefore, if anyone is in Christ, the new creation has come: The old has gone, the new is here!*

When we follow Him, Christ makes us a new creation, and we can tell others the story of how God has changed us. The story of how God has changed someone is called their testimony. We all have a testimony to share as we grow in our relationship with Christ. Hey, Hannah, let's ask Uncle Gus to share his testimony with us the next time we see him.

How has God changed your life or the life of someone in your family? Ask someone to share their testimony with you. Look up 2 Corinthians 5:17. If you'd like, you can highlight this verse in your Bible. Memorize 2 Corinthians 5:17 with your teacher or with a sibling.

I hope you brought your Science Notebook and art supplies with you today, friend. We're going to add a new page to our Science Notebook!

Let's draw a picture of some sea stars! I have a picture of some in a tide pool right here. You can draw your sea stars in different patterns and colors if you'd like!

Here is what each of our Notebook drawings look like. I drew one orange sea star with big arms. Sam decided to draw a little red sea star, isn't it cute? Hannah drew three sea stars with different patterns — they look really neat! I hope you have as much fun drawing your sea stars as we did, friend.

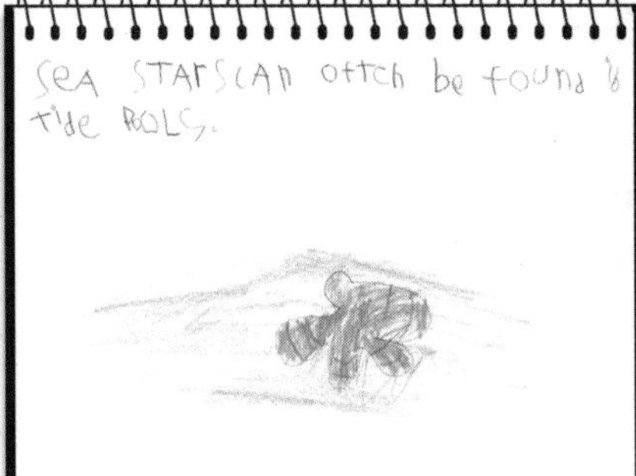

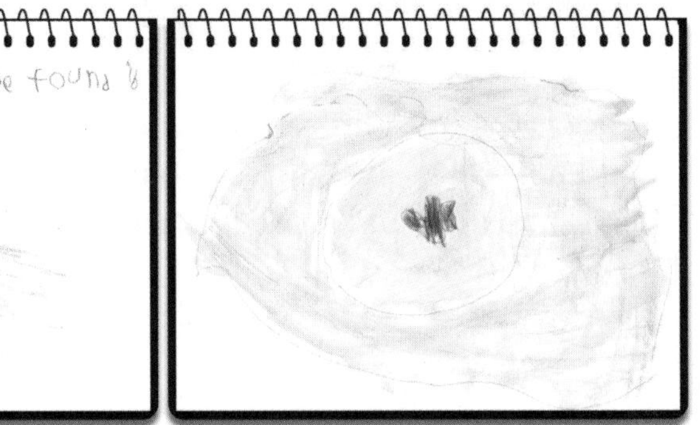

In your Notebook, write:

Sea stars can often be found in tide pools.

Then draw a picture of sea stars in a tide pool.

Learning about sea stars this week also reminded us that God gives us new life in Him. Copy 2 Corinthians 5:17 on the back of your Notebook page as a reminder.

*Therefore, if anyone is in Christ, the new creation has come: The old has gone, the new is here!* (2 Corinthians 5:17).

# Tide Pools 2

Ahoy, friend, I'm glad you're back to continue learning about tide pools with us. Our mom told us that we would be creating our own model of a tide pool later this week — I'm so excited to get started!

**Note**

The student will complete several projects during the course of this week. For a full list of required materials, please refer to the Master Materials List.

**materials needed**

- Air dry modeling clay
- Purple acrylic paint
- Paintbrush
- Paper plates
- Plastic tablecloth

Last week, we learned how tide pools are created and about sea stars. Do you remember what phylum sea stars are classified in, friend? They are echinoderms. Let's learn about two other types of echinoderms this week: sea urchins and sea cucumbers.

Like sea stars, sea urchins and sea cucumbers can be found in oceans all around the world. They are also among the living things we might spot in a tide pool. Now, you may have guessed from the name that sea cucumbers have a body that looks a bit like a cucumber — and you'd be right!

There are over 1,000 species of sea cucumbers found all around the world. One common species is called the hairy sea cucumber. Hairy sea cucumbers are found in the Atlantic Ocean along the coast of North America, the Gulf of Mexico, and even the Caribbean. The hairy sea cucumber can grow to be five inches long. This species is typically a plain brown color, though other species of sea cucumber have more vibrant blues, greens, or even red.

Let's talk about sea urchins now. These spiny creatures remind me of a pincushion!

The purple sea urchin can be found along the coast of the eastern Pacific Ocean, and they can grow to be around three inches across. Like sea stars, sea urchins use their tube-like feet to move around. But that's not all their feet can do! At the end of their feet is a feature like a suction cup that they can use to secure themselves to their surroundings. This design keeps them from being washed away in stronger waves or tides.

Neat. Now we know that God was the original creator of the suction cup!

A sea urchin is covered in protective spines. When it senses danger, it can point its spines toward the threat to help protect itself. With those protective spines, you might think that a sea urchin would have no predators, but that's actually not the case. Seagulls, sea otters, and even some types of sea stars aren't afraid of those purple spikes, and they will eat a sea urchin. We'll talk more about the ocean's food chain in a couple weeks.

Hey — I have an idea in the meantime. Let's create our own model of a purple sea urchin today. Be careful as you create your model — you don't want to be poked by the spines!

## Activity directions:

1. Spread out a plastic tablecloth to protect the surface you're working on.
2. Roll out two small balls from the modeling clay. They should each be about 1 to 2 inches across.
3. Carefully insert toothpicks into the balls to create the sea urchin's spines.

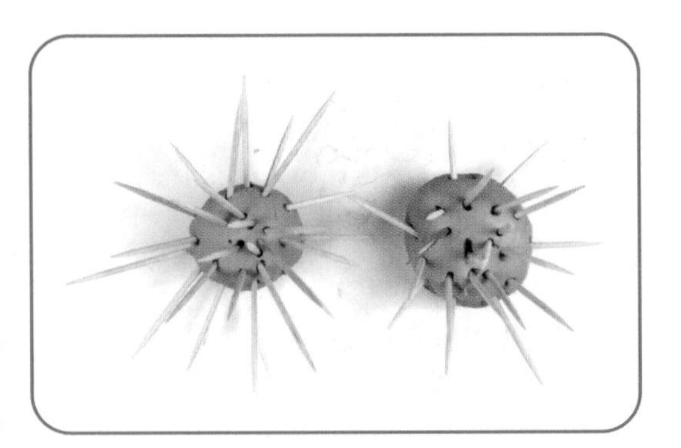

4. Place the sea urchins on a paper plate and allow them to dry. Watch out for the spines! Depending on how thick the clay is, it may take 1–3 days to dry.
5. Once the clay has dried, spread out the plastic tablecloth to protect your surface. Carefully paint the sea urchins and the spines purple.

6. After the paint has dried, show someone your model sea urchins and share what you've learned about them. Then place your sea urchins in a safe place. We'll be using them again later this week.

Welcome, friend! We're continuing to learn about the living things we may be able to spot along the ocean's shore in a tide pool. We've been learning about echinoderms together, but it's time to learn about a new group of living things today.

That's right, we're going to learn about mollusks (said this way: mŏl-ŭhsks)! Mollusks belong to the Mollusca (said this way: mŭh-lŭhs-kŭh) phylum. Mollusks include living things such as clams, mussels, oysters, snails, squid, and even octopuses.

**materials needed**

- Paper plates
- Gray or brown acrylic paint
- Paintbrush
- Stapler
- Plastic tablecloth
- Scissors

If you've ever been to the ocean, you may have found part of a shell from a clam, mussel, or oyster. I wonder, what are their shells made of?

Ooh, good question! A mollusk's shell is typically made from a compound called calcium carbonate (said this way: kăl-sē-ŭhm kår-bŭh-nāt). Remember, a compound is formed when two or more different atoms bond together to form a molecule. The chemical formula for calcium carbonate is $CaCO_3$. Do you remember how to read a chemical formula, friend? Let's do it together!

First, we'll need to look at the periodic table of elements — go ahead and grab your copy of the periodic table, friend! What element is Ca? It is calcium! We also know that C is the symbol for carbon and O is the symbol for oxygen. This means that a molecule of calcium carbonate is made from 1 atom of calcium, 1 atom of carbon, and 3 atoms of oxygen.

Neat! It's a good thing we learned about chemistry earlier this year. Let's learn a little about clams together today. There are many thousands of species of clams that live in oceans all around the world. Some species of clam can even be found in fresh water.

A clam's shell is made from two halves, the upper and lower shell. These two halves are joined together by a ligament. A ligament helps to hold a joint, like an elbow, knee, or shell, together. You have ligaments in your body too! The clam's ligament works like the hinge on a door. It holds the shell together but also allows it to open and close. How does a clam move around, Ben?

Well, I have two feet, and sea stars have many tube-like feet, but a clam only has one foot. A clam can stick its muscular foot outside of its shell to shuffle itself along. But a clam's foot is most useful for digging so that the clam can burrow itself beneath the sand. Can you imagine being able to dig a great hole with just one foot?

That would be an awful tough job — but God gave clams the perfect design to be able to do it. Let's create a few clams of our own today!

## Activity directions:

1. Spread out a plastic tablecloth to protect the surface you're working on.
2. Place two paper plates on top of each other. Start at the rippled edge of the plate and cut an oval shape between 1–3 inches wide.
3. Separate the plates — these are your two clam shell halves. Flip one half upside down and align the edges. Staple the back edge together. The staple will work like the clam's ligament to allow the shell to open and close.

ASK PARENT FOR HELP

4. Repeat this process to create 3–4 clams.
5. Paint your clams brown or gray. Allow to dry then set them in a safe place. We'll be using them later this week!

Are you ready to continue exploring tide pools with us, friend? We are going to learn about another creature that can often be found hiding in tide pools — crabs!

There are well over 6,000 different species of crabs found in fresh and salt water all around the world. Today, we're going to learn more about a particular species of crab, the striped shore crab.

The striped shore crab can be found around the coast of the Pacific Ocean along the continent of North America. They can be spotted in and around tide pools — but you may have to look closely! Their dark, striped coloring helps to camouflage them in the dark, rocky spaces of tide pools and the ocean.

**materials needed**

- Paper plate
- Black & green acrylic paint
- Hot glue gun (adult only)
- Plastic tablecloth
- Scissors
- 3 red pipe cleaners
- 2 googly eyes

**Camouflage** (said this way: kă-mŭh-flåzh) is a way to stay hidden in an environment. God gave many living things different colors or patterns that would help them stay camouflaged in their natural habitat.

The striped shore crab grows to be just about 1–2 inches wide. It is quite a small crab. It spends about half of its time on land and half in the water. Striped shore crabs primarily eat algae — we'll be learning more about algae next week. However, they'll also eat snails, worms, mussels, or even fish that have died.

Crabs are invertebrates, but their hard shell functions as a skeleton on the outside of their body. This is called an **exoskeleton** (said this way: ĕk-sō-skĕl-ĭ-tŏn). The crab's exoskeleton supports and protects its body underneath.

Hmm, interesting! But how can a crab grow with that hard shell on the outside of its body? Does the shell grow, like our bones do?

Great question, Ben. No, the crab's exoskeleton doesn't grow with it like our bones grow with us. Instead, God designed the striped shore crab with a soft, new exoskeleton underneath the now-too-small one. Once the crab has outgrown the outer exoskeleton, it will shed the old shell by crawling out of a small slit in the back of it. The crab will then leave the old exoskeleton behind. This is called **molting** (said this way: mōlt-ing). Once the crab has molted, the new exoskeleton that was underneath the old one begins to slowly harden.

What a neat design God gave them! Do any other living things molt like the crab?

Yup! Crabs belong to the Arthropoda phylum. Arthropods (said this way: år-thrŭh-pŏds) include crabs, lobsters, spiders, and insects. One feature these living things have in common is molting.

Now I have a surprise for you — we're going to create our own model of a striped shore crab! Let's get started.

## Activity directions:

1. Spread out a plastic tablecloth to protect the surface you're working on.
2. Cut the shape of the crab's body from the paper plate — the rippled edge should be slightly wider than the back edge of the body. The body should be about 3–4 inches wide at the front. This will give you enough room to complete the craft.

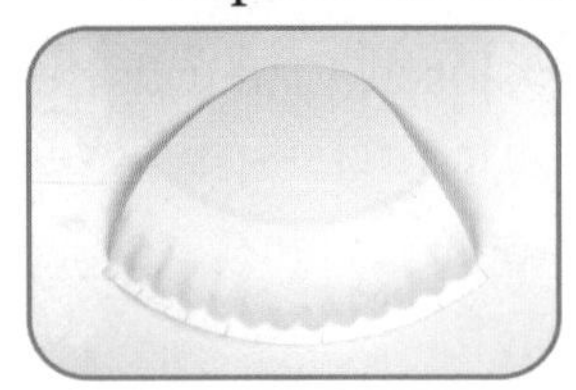

3. Cut one pipe cleaner into four equal pieces then repeat with the second pipe cleaner. These will be the crab's legs.
4. Cut the third pipe cleaner in half. Form an 'M' shape at one end of each pipe cleaner. Leave one of the legs of the 'M' a bit longer than the other. These will form the crab's claws.

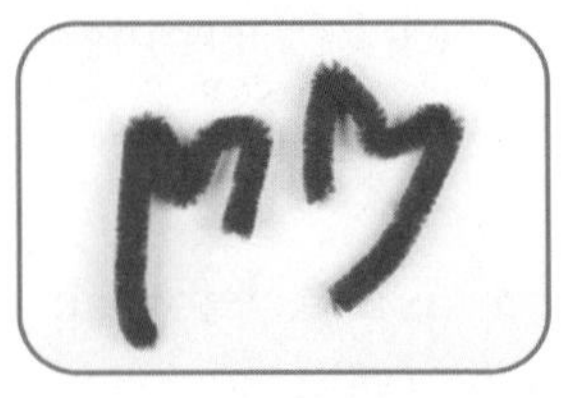

5. Paint the crab's body black and allow it to dry. Once it has dried, add green stripes to the shell.
6. After the crab's shell has dried, ask your teacher to use the hot glue gun to attach the legs to the bottom of the crab's body. There should be four pipe cleaner legs on each side. Once the glue has dried, you can bend the legs so that the crab can stand on them.
7. Ask your teacher to pinch one of the crab claw 'M's together and use a bead of hot glue to attach the shorter side of the 'M' shape to the middle of the longer side. Repeat with the other 'M.' Then attach the claws to the front of the crab.
8. Once the glue has dried, attach googly eyes to the crab.

9. Show someone your striped shore crab and tell them what you learned! Then place it in a safe place. We'll be using it later this week.

Hey there, friend! I hope you had fun this week learning a little about some of the living things we may find in a tide pool. We're going to create our own model of a tide pool tomorrow using the crafts we've been working on during the last two weeks. I can hardly wait! But in the meantime, I've been thinking about the striped shore crab and the ways God gave it to defend itself.

You mean its hard exoskeleton, camouflage, and claws?

Exactly! Its exoskeleton protects its soft body underneath, like armor. The crab's camouflage helps it to hide and disappear from the sight of anything that might want to harm it. The crab can also use its claws to scare away or pinch something that threatens it. I think it is amazing that God designed creation perfectly at the beginning — but He also gave many living things the features they would need to survive in a world broken by sin.

I see God's wisdom and mercy in that too. Did you know that just like God equipped the crab with what it would need in a broken world, He also equips us with what we need to defend against the lies of the enemy? The Bible compares our enemy, the devil, to a roaring, dangerous lion. The devil tries to lie and deceive us, just as he did to Adam and Eve in the Garden of Eden.

In Ephesians 6:10–11, the Apostle Paul wrote,

*Finally, be strong in the Lord and in his mighty power. Put on the full armor of God, so that you can take your stand against the devil's schemes.*

In Ephesians 6:14–17, Paul goes on to describe the armor God gives us to help us stand firm against the enemy:

*Stand firm then, with the belt of truth buckled around your waist, with the breastplate of righteousness in place, and with your feet fitted with the readiness that comes from the gospel of peace. In addition to all this, take up the shield of faith, with which you can extinguish all the flaming arrows of the evil one. Take the helmet of salvation and the sword of the Spirit, which is the word of God.*

Now, we don't walk around wearing a real suit of armor like a soldier. Instead, Paul wrote these verses to give us a picture of how the Word of God and His truth protects our hearts and minds from the lies of the enemy.

Right! We continue to study God's Word and to make sure it is the foundation of our worldview so that we can continue to stand firm in our faith with the armor of God that Paul described.

Ask a friend or family member how they have stood firm in their faith. Then look up Ephesians 6:10–11 in your Bible. If you'd like, you can highlight these verses. Memorize Ephesians 6:10–11 with your teacher or with a sibling.

Hey there, friend, I'm glad you're here! Usually, this is the day we add a new page to our Science Notebook, but we're going to do something a bit different today!

Yes, we're going to create our own model of a tide pool together. We've already created models of a sea star, sea urchins, clams, and a striped shore crab. Now we just need to create a model of a tide pool to put them in!

I've got a shoebox here. We can use some construction paper to create the bottom of our tide pool. Then we can add some rocks or sand to create the edge of the tide pool. Once our tide pool is ready, we'll add our model creatures.

This is going to be so much fun! Let's get started.

**materials needed**

- [ ] Shoebox
- [ ] Rocks or sand
- [ ] Blue construction paper
- [ ] Plastic tablecloth
- [ ] Scissors
- [ ] Craft model sea star, sea urchins, clams, and crab from previous lessons
- [ ] Acrylic paint & paintbrush

## Activity directions:

1. Spread out a plastic tablecloth to protect the surface you're working on.
2. If you'd like, you can paint the ocean or the sky on the back of the shoebox. Allow the paint to dry.
3. Cut the construction paper so that it fits inside the bottom of your tide pool model.
4. Use rocks or sand to create the edge of your tide pool.
5. Carefully add your model sea star, sea urchins, clams, and crab to the model tide pool.

6. Share your tide pool model with your family. Be sure to tell them about the creatures you learned about, as well as the special designs God gave them.

**Bonus!** Take a picture of your tide pool model and ask your teacher to help you print it out. Then tape or glue the picture on the next page in your Science Notebook. Write **My Tide Pool Model** at the top of the page.

week 24

# Algae & Seagrass

Hey there, friend! Are you ready to begin another science adventure this week? We're going to be exploring meadows and forests.

Wait, meadows and forests? I thought we were learning about marine biology! When did our topic suddenly change?

Oh, don't worry — we're still studying marine biology. I'm talking about marine meadows and forests.

Ah, you're a tricky one, Hannah! I didn't know there were meadows and forests in the ocean.

There are! The ocean is home to living things such as algae, seaweed, seagrass, and kelp. We're going to talk about them this week as well as learn about seagrass meadows and kelp forests.

Cool, let's start with algae (said this way: ăl-jē). I had to clean my fish tank this week because the glass was getting covered in algae. I'm curious, what is algae anyway? Is it a type of plant?

Though algae does photosynthesize like a plant, it isn't actually a plant. Do you remember when we talked about taxonomy and the classification system a few weeks ago, Ben?

Sure do! We learned about two kingdoms in taxonomy, the plant and animal kingdoms. If algae isn't a plant, does that mean it's an animal?

Nope! Algae isn't a plant or an animal. But don't worry — the plant and animal kingdoms aren't the only kingdoms in taxonomy. The third kingdom in taxonomy is called Protista (said this way: prōh-tĭs-tŭh). We call organisms in this kingdom protists. Protists are living things that aren't animals, plants, bacteria, or even fungi. Algae is classified in the Protista kingdom.

Ah, so algae is a rather unique kind of living thing.

It sure is. Algae is found all over the world in ponds, streams, lakes, the ocean, and even your fish tank. Algae can grow in fresh or saltwater environments. You may also find algae growing at times on rocks, tree trunks, or an animal's fur — like the sloth!

Name: ______________________

There are many thousands of species of algae. We often group them by their color: green, red, blue, or brown. Though it may look simple, algae is incredibly important to life on earth.

Well, I can say that algae is sure good at coating the glass inside my fish tank, but that doesn't seem like an important job. What makes algae so important?

Algae is responsible for producing most of the oxygen on the earth — even more than all the plants on earth produce. That is a pretty important job, and it makes algae very important to all living things. We'll be learning more about algae as we continue exploring this week!

apply it

1. What are the three kingdoms in taxonomy that we've learned about?

2. What kingdom is algae classified in?

3. What makes algae important to life on earth?

4. Have you ever seen algae growing in a fish tank, pond, lake, or on a rock or tree?

5. If you have, what did it look like?

Whew, I just finished helping Mom pull weeds out of the vegetable garden. It was tough work! I'm glad it's done now, and I'm ready to start our science adventure for today. What are we going to learn about, Hannah?

We're going to learn about seaweed today.

Oh, it must be the day for weeds.

Well, unlike the weeds in the vegetable garden, seaweed isn't actually a plant. Seaweed is a type of algae.

That means seaweed also belongs to the Protista kingdom, then?

Yup! There are thousands of species of seaweed all over the world. Like simple algae, we can group these species by their color: green, brown, or red. Green algae is typically found in warm, tropical oceans.

Imagine the earth for a moment, then imagine drawing a horizontal line around the very middle of it. This imaginary line is called the equator. The word tropical refers to the region of the earth that is around the equator. The tropical region is usually warm and humid all throughout the year.

Thanks, Ben! Brown seaweed prefers to grow deep in the cooler waters of the northern hemisphere. The northern hemisphere is the top half of the earth above the equator. Red seaweed also prefers to grow in cold water.

Now, remember, seaweed is not a type of plant. It does not have a root system to draw in nutrients from the soil as a plant would. Instead, seaweed uses sunlight to photosynthesize and also draws in nutrients to itself from the water around it.

Interesting! But I have a question now. Since it doesn't have roots to hold itself in place, does seaweed just float along in the ocean?

Though you may occasionally find clumps of seaweed on shore or floating in the ocean, seaweed uses a root-like structure called a holdfast to anchor itself in place. Holdfasts do not absorb nutrients like plant roots — their only job is to hold the seaweed securely in place.

In the ocean, seaweed provides a habitat for other living things and is also an important part of the food chain. We'll be talking more about the marine food chain next week!

Name: ______________________________

God created many different species of seaweed. Let's look at a few!

What is something interesting you notice about seaweed?

______________________________

______________________________

______________________________

______________________________

______________________________

______________________________

Hello there, friend! Are you ready to learn about more algae today?

Actually, I'm kind of ready to explore marine forests and meadows.

Well, then let's dive right in! Kelp is a type of seaweed found along the North American coast of the Pacific Ocean. Kelp can grow quite tall, and very quickly. In fact, some species of kelp can grow up to 18 inches in a single day!

Whoa — that's a lot of growing!

Indeed. Kelp looks a bit like a leafy plant, and it has three basic parts. The first part is the holdfast. Remember, a holdfast is a root-like structure that holds seaweed firmly in place. Kelp can grow to be up to 175 feet tall, so it definitely needs a secure holdfast! The second part of kelp is the sturdy stalk that allows it to grow tall, and it is called the stipe. Finally, blades grow from the stipe. Kelp blades are a bit like the leaves on a plant, and they photosynthesize sunlight that filters through the water.

Several species of kelp can grow together in an area, and their quick, tall growth creates what we call a kelp forest. Kelp forests provide an important marine ecosystem that is home to many different living things, including vertebrates like fish, turtles, and sea otters, and invertebrates like sea stars and sea urchins.

That is really cool! I'd like to go scuba diving in a kelp forest sometime. You also mentioned meadows, Hannah. What creates an underwater meadow?

Ocean meadows are formed by seagrass. Seagrass is not a type of algae; it is a plant with roots, leaves, and often flowers. Seagrass can usually be found in warm, shallow tropical waters. Seven species of seagrass are found around the coast of Florida in the United States. These seven species are named shoal grass, manatee grass, Johnson's seagrass, turtle grass, paddle grass, widgeon grass, and star grass.

Grassy meadows on land provide an important habitat and food source for other living things, and seagrass meadows are no different. Seagrass roots help to hold dirt, sand, rocks, and nutrients in place along the ocean floor. They also produce oxygen through photosynthesis, filter debris in the water, provide a place to hide for small marine creatures, and are an important source of food.

Name:

Copy the name of each type of seagrass.

Shoal grass

Manatee grass

Turtle grass

Paddle grass

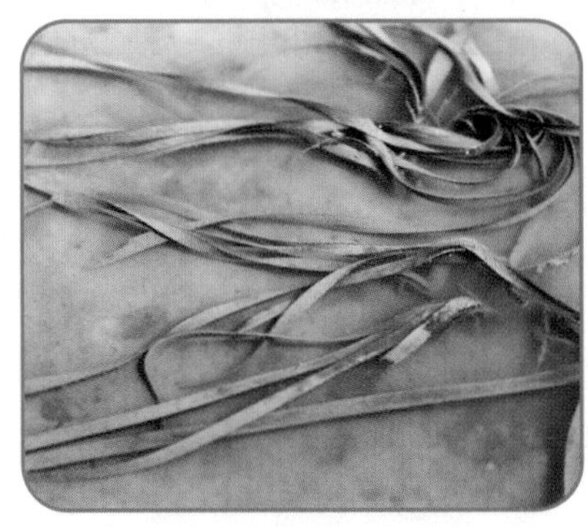

Widgeon grass

Star grass

I don't know about you, but I've sure enjoyed learning more about algae and seagrass this week. Someday, I hope I can see a kelp forest and a seagrass meadow in the ocean for myself!

That would be a lot of fun!

As we've been learning this week, I've been thinking about holdfasts. Without holdfasts, seaweed would be tossed back and forth by the waves, like leaves blown all around by the wind. It wouldn't be secure — it would just float around wherever the ocean waves and tides pushed it.

Sometimes, the world around us can feel like the waves and tides of the ocean. It tries to change what truth is and pushes us in all different directions. We've been talking this year about how the Bible is the firm foundation that we build our worldview on. We turn to God's Word in the Bible to tell us what truth is — and we know that the truth in God's Word does not ever change.

Just as seaweed's holdfast keeps it secure against the waves, the Bible is our holdfast. God's Word holds us secure against the lies of the enemy in the world around us.

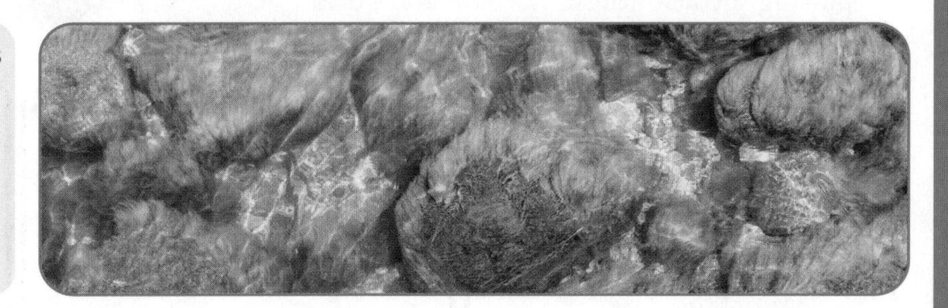

In Ephesians 4, the Apostle Paul talks about how we become mature, or full-grown and strong, in our faith. In Ephesians 4:14 he wrote,

*Then we will no longer be infants, tossed back and forth by the waves, and blown here and there by every wind of teaching and by the cunning and craftiness of people in their deceitful scheming.*

Dad and Mom have been talking to us about how there are many different teachings in the world around us — but not all of them are based on what the Bible tells us. It is important that we learn what God's Word says; this is how the Bible becomes the foundation of our worldview. When we know what God tells us in His Word, we are able to recognize false teachings and the lies of the devil.

In other words, when we know what God tells us in the Bible, it is our holdfast! It keeps us secure so that we are not tossed and blown about by things that are not the truth revealed in the Bible.

God's Word is secure and unchanging. Talk to your family about ways you can learn to compare the things you see, read, or hear to what the Bible says. Look up Ephesians 4:14 in your Bible. If you'd like, you can highlight this verse. Memorize Ephesians 4:14 with your teacher or with a sibling.

Hey, Ben, do you know what day it is?

Oh, Hannah, you're so funny . . . do I know what day it is? It's the day we get to add a new page to our Science Notebook! I've been looking forward to today all week long, and I'm ready to go.

Well, let's get started then! We learned about seaweed and seagrass this week, so I thought it would be fun to draw a picture of a kelp forest in our Science Notebook. I have a picture right here that we can use to give us an idea for our own drawings.

Here is how each of our drawings turned out. If you'd like, friend, you can even add a fish or scuba diver exploring the kelp forest like Sam and I did!

In your Notebook, write:

Kelp forests provide an important marine ecosystem.

Then draw a picture of a kelp forest.

Learning about seaweed's holdfasts this week also reminded us that God's Word is what holds us secure. Copy Ephesians 4:14 on the back of your Notebook page as a reminder.

*Then we will no longer be infants, tossed back and forth by the waves, and blown here and there by every wind of teaching and by the cunning and craftiness of people in their deceitful scheming* (Ephesians 4:14).

week 25

# Marine Food Chain

Day 1

Hey there, friend, welcome to our next science adventure! This week, we're going to explore the marine food chain together. The food chain is all about the links between plants, animals, and people as animals and people eat. The first big question we need to answer is, why do animals and people need to eat anyway?

The answer to that question is energy. Energy is what allows us to work, play, learn, and live. All living things need energy in order to survive. When we eat, we receive energy from the food, and our body uses that energy to help us do what we need to do.

Thanks, Ben! Now, before we go too much further, let's look at God's original design for the food chain. Let's read Genesis 1:29–30 together and see what we can learn from these verses.

*Then God said, "I give you every seed-bearing plant on the face of the whole earth and every tree that has fruit with seed in it. They will be yours for food. And to all the beasts of the earth and all the birds in the sky and all the creatures that move along the ground — everything that has the breath of life in it — I give every green plant for food." And it was so.*

Hmm, the first thing I noticed is that God designed animals and people to be able to eat. Food was necessary, even at the very beginning!

I also noticed that in God's original creation, He gave both animals and people the plants for food. The food chain would have been pretty simple and short: it began with plants and ended with the animal or person that ate the plant!

But this isn't the food chain that we observe today. Today, we may notice a mouse eat a flower in a meadow, then a snake comes along and eats the mouse. Later that week, a hawk swoops down and eats the snake. So now the big question is, what changed? What happened to the original food chain?

- 6 Styrofoam™ cups ☐
- Glue ☐
- Scissors ☐
- Permanent marker ☐

**Weekly materials list**

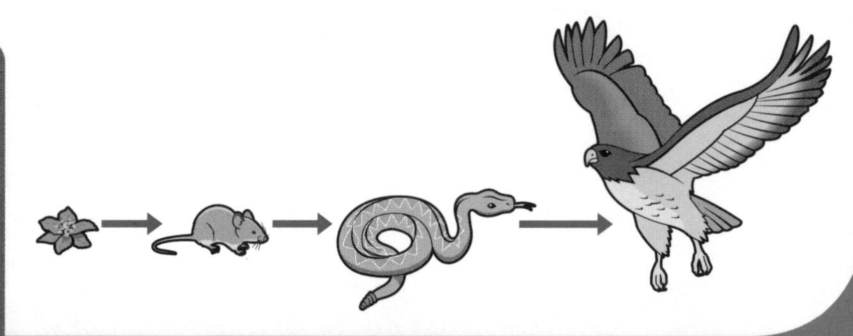

Name: ____________________

Sin entered creation. Sin changed things, creation became imperfect, and death entered creation as a consequence of sin. Eventually, creatures would have begun to eat other creatures, forming the food chain we observe today. After the Flood in Noah's time, God also gave people permission to eat meat in Genesis 9:3.

*Everything that lives and moves about will be food for you. Just as I gave you the green plants, I now give you everything.*

The Bible explains the food chain that we see today right in the Book of Genesis! We'll continue to talk more about the marine food chain in our adventures this week.

apply it

1. What do you think it would have been like to live in God's original creation where animals and people ate only plants?

2. Look up Genesis 9:1–2 in your Bible. What did God tell Noah would happen to the animals after the Flood?

3. What do you think it would have been like to see a lion, tiger, bear, or dinosaur before it was afraid of people?

Welcome back! We're exploring the marine food chain together this week. In our last adventure, we talked about God's original design for the food chain, as well as why the food chain is different today. Now that we've laid our foundation, it's time to explore the food chain deeper.

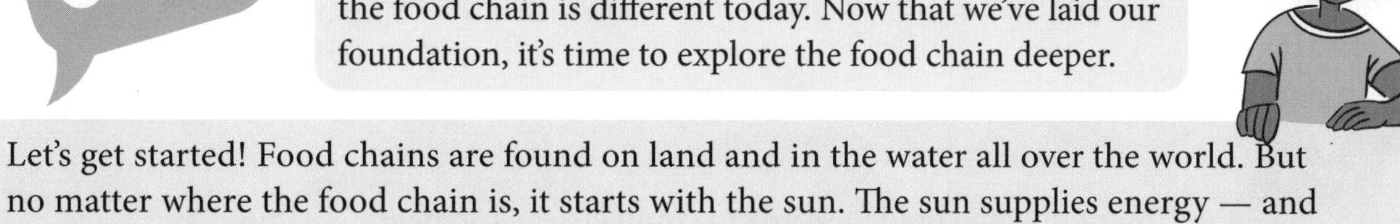

Let's get started! Food chains are found on land and in the water all over the world. But no matter where the food chain is, it starts with the sun. The sun supplies energy — and remember, the food chain is all about how energy moves through creation from one living thing to another.

Plants and algae need energy to grow and live. But, unlike animals and people, plants don't eat anything to receive energy. Instead, they produce, or make, their own food through photosynthesis, using energy from the sun. That's why we call plants and algae **primary producers** in the food chain. Primary producers produce their own food for energy. In the ocean, algae, seaweed, and seagrass would be examples of primary producers.

Another important type of primary producer in the ocean is called phytoplankton (said this way: fī-tō-plănk-tŏn). Phytoplankton is an extremely tiny type of algae. In fact, you usually need a microscope to be able to see it! Phytoplankton, algae, and seagrass are all primary producers and are the base of the marine food chain.

The next link of the food chain is called consumers. A **consumer** (said this way: kŭn-soo-mer) isn't able to make its own food like a primary producer can. Instead, it needs to consume food from a plant or animal in order to receive energy. Consumers can be organized into three groups: herbivores, carnivores, and omnivores.

An **herbivore** (said this way: er-bĭh-vōr) is a consumer that only eats primary producers like plants and algae. Zooplankton (said this way: zoo-plănk-tŏn) are marine animals that eat phytoplankton, which makes them herbivores. Like phytoplankton, zooplankton are also only visible under a microscope.

The next group of consumers are called carnivores. A **carnivore** (said this way: kår-nĕ-vōr) is a consumer that only eats other consumers like fish or animals. A shark might be the first thing that comes to mind when you think about ocean carnivores!

Finally, an **omnivore** (said this way: åm-nĭh-vōr) is a consumer that eats both producers and other consumers. Many species of fish, crabs, and even sea urchins are omnivores.

**Zooplankton (herbivore)**

**Shark (carnivore)**

**Tuna (omnivore)**

Name: ______________________________

Whew! We covered a lot today — let's review what we've learned. I can't wait to continue learning about the food chain in our next lesson.

Unscramble the bolded words to complete each sentence.

1. Plants and algae are examples of **pyiarmr cpordures**.

______________________________

2. A consumer that eats both producers and other consumers is called an **ionmveor**.

______________________________

3. A consumer that eats primary producers is called an **oreverbih**.

______________________________

4. A consumer that only eats other consumers is called a **narivorec**.

______________________________

Hey there, friend! Do you remember what we talked about last time we were together? We learned about producers and consumers. We also learned how consumers can be organized into three groups: herbivores, carnivores, and omnivores. Today, let's use what we've learned to construct a marine food chain.

**materials needed**

- 6 Styrofoam™ cups
- Glue
- Scissors
- Permanent marker

The marine food chain begins with the sun. The sun provides energy to primary producers like algae, seaweed, phytoplankton, and seagrass so that they can photosynthesize. What comes next, Ben?

Consumers are next in the food chain. The first group of consumers is called primary consumers. Primary consumers are herbivores — they only eat primary producers like plants or algae. In the ocean, zooplankton are an example of a primary consumer because they eat phytoplankton. Some types of small fish and shrimp are also primary consumers.

Alright, so our food chain so far is made of the sun, a primary producer like phytoplankton, and a primary consumer like zooplankton. The next link in the food chain is called secondary consumers. Secondary consumers can be omnivores or carnivores. Fish, sea urchins, crabs, sea otters, sea stars, and some species of whale are all examples of secondary consumers.

Secondary consumers aren't at the end of the food chain because they can also be the prey of a bigger predator.

Wait a minute, let's make sure we know what those words mean! **Prey** (said this way: prāy) means an animal that is hunted and eaten by another animal. A **predator** (said this way: prĕ-dŭh-ter) is a consumer that hunts and eats other animals.

Right! Thanks, Ben. The next link in the food chain is called tertiary consumers. Tertiary (said this way: ter-shē-āir-ē) is a word that means third. These are the third group of consumers in the food chain. In a marine food chain, a few examples of tertiary consumers are large fish, sea lions, and turtles.

Finally, the apex predators are at the very top of the food chain. **Apex** (said this way: āpĕx) means the top. So, an apex predator is a predator that is at the very top of the food chain. You might think of large sharks or the orca whale as apex predators in the ocean — and you'd be right! Apex predators are consumers that are typically not the prey of any other predator.

We've learned a lot of information today. Hey, I have an idea — let's create a marine food chain to help us remember what we've learned!

## Activity directions:

ASK PARENT FOR HELP

1. Hold the first Styrofoam™ cup upside down. On the rim of the cup, write "Sun." Then turn it around and write "Energy" on the back of the rim. Cut out the picture of the sun from the worksheet and glue it to the cup, above the "Sun" label.
2. Hold the second cup upside down. Write "Phytoplankton" on the rim. Then write "Primary Producer" on the back of the rim. Cut out the picture of phytoplankton and glue it to the cup, above the "Phytoplankton" label.
3. Hold the third cup upside down. Write "Zooplankton" on the rim. Then write "Primary Consumer" on the back of the rim. Cut out the picture of zooplankton and glue it to the cup, above the "Zooplankton" label.
4. Hold the fourth cup upside down. Write "Fish" on the rim of the cup. Then write "Secondary Consumer" on the back of the rim. Cut out the picture of the fish and glue it to the cup, above the "Fish" label.
5. Hold the fifth cup upside down. Write "Sea Lion" on the rim. Then write "Tertiary Consumer" on the back of the rim. Cut out the picture of the sea lion and glue it to the cup, above the "Sea Lion" label.
6. Hold the sixth cup upside down. Write "Orca" on the rim. Then write "Apex Predator" on the back of the rim. Cut out the picture of the orca and glue it to the cup, above the "Orca" label.
7. Allow the glue to dry. Once dried, place the "Sun" cup on the table with the rim down. Next, stack the phytoplankton cup on top. Repeat with each cup to create a marine food chain. Show someone your marine food chain and share what you've learned this week.

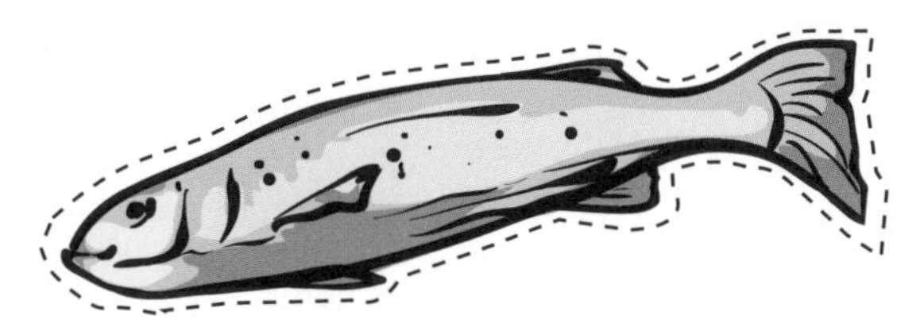

Fish

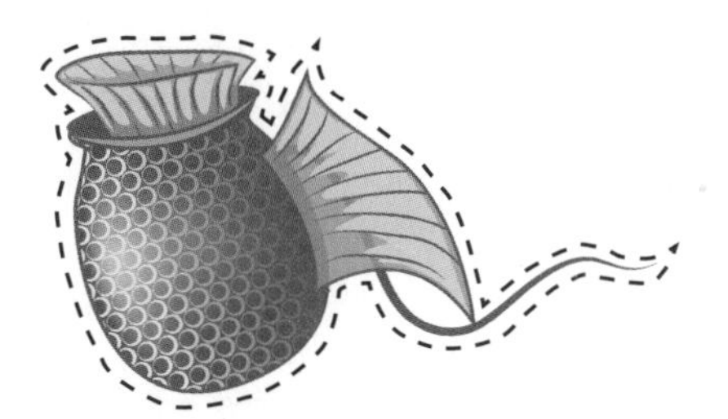

Phytoplankton

Zooplankton

Sea Lion

Orca

Blank for cutting purposes.

Whew, we're coming to the end of this week's adventure. There is definitely a lot to explore in the food chain. We learned about a simple marine food chain this week — but there are many different food chains all around the world!

Very true! The food chain that we see today isn't the food chain that God originally designed for creation — but did you know that the food chain we observe today is still important? As we talked about earlier this week, we know from reading Genesis that God's original creation was perfect — there was no sickness or death. In God's perfect design, all consumers ate only plants.

However, once sin entered creation, things changed. Sickness and death entered creation as a consequence of sin, and the original design was broken. But we can still see God's wisdom and care, even in the broken world around us. In this broken creation, things can quickly become imbalanced. The food chain that we observe today helps to keep creation in balance.

For example, do you remember what we learned about purple sea urchins a couple of lessons ago? These spiny creatures love to eat kelp in a kelp forest. But sea urchins can also quickly destroy an entire kelp forest when there are too many urchins in an area. If a kelp forest is destroyed, it also destroys the habitat of hundreds of other marine creatures. Soon, all we might find in an area would be purple sea urchins rather than all the different, beautiful marine creatures living together. That would be so sad!

Ah, I understand. Predators, like sea otters, eat purple sea urchins and, by doing so, help to keep the number of sea urchins balanced. The food chain helps to protect the kelp forest and preserve the habitat of many diverse marine creatures.

Right! The food chain we observe today is a sad reminder of the consequences of sin — but it is also a reminder of God's wisdom and mercy even in a broken creation. The food chain helps to keep habitats and ecosystems balanced so that we can enjoy watching and learning about many of God's creations in one place. It reminds me of Psalm 104:24–25,

*How many are your works, LORD! In wisdom you made them all; the earth is full of your creatures. There is the sea, vast and spacious, teeming with creatures beyond number — living things both large and small.*

I'm glad that God designed all things in His wisdom and that the food chain helps to keep the sea teeming with living things, both large and small, for us to learn about!

Look up Psalm 104:24–25 in your Bible. If you'd like, you can highlight these verses. Memorize Psalm 104:24–25 with your teacher or with a sibling.

Hey there, friend! We're going to add a new page to our Science Notebook today.

Oh, oh, oh! I have an idea for this week — can we draw a picture of a sea otter?

I like that idea. We learned about the food chain this week as well as how predators, like the sea otter, help to keep creation balanced so that we can enjoy seeing may diverse creatures. Here is a picture we can use for an example to help us draw our own sea otters.

Here is how each of our sea otters turned out. We can't wait to see yours, friend!

In your Notebook, write:

Predators, like the sea otter, help to keep marine ecosystems balanced.

Then draw a picture of a sea otter.

Learning about the food chain this week also reminded us that it keeps a broken creation balanced so that we can enjoy seeing many of God's creations together in habitats and ecosystems. Copy Psalm 104:25 on the back of your Notebook page as a reminder.

*There is the sea, vast and spacious, teeming with creatures beyond number — living things both large and small* (Psalm 104:25).

week 26

# The Coral Reef

Day 1

Hello, friend! Are you ready to explore a new marine ecosystem together this week? An ecosystem is a community of living and nonliving things that are together in one place. God designed many living things to be dependent on one another in an ecosystem. We're going to begin exploring the coral reef and some of the relationships God designed for this ecosystem next.

Ooh, this will be fun! I've seen pictures of coral reefs in my books about the ocean. Coral reefs are some of the most beautiful and diverse ecosystems on the earth.

They are, indeed. Coral reefs grow in warm, tropical waters around the equator. We can organize coral reefs into three different types. The first type of coral reef is called a barrier reef.

Hmm, the word barrier means something that creates a division or obstacle, like a fence. Does this type of reef create a barrier between the land and ocean?

Good guess, but no. Before we answer that question, we need to learn a new word. A **lagoon** (said this way: lŭh-goon) is an area of still, quiet ocean water that is separated from the ocean by a sandbank or island.

A barrier reef is separated from the shore by deeper water or a lagoon. The water between the shore and the reef creates a barrier — that is an easy way to remember what makes a barrier reef! Though it is separated from the shore, a barrier reef runs along or parallel to the shoreline. One of the most well-known barrier reefs in the world is called the Great Barrier Reef. This amazing coral reef is found off the shore of Australia. We're going to learn more about it next week!

The second type of coral reef is called a fringing reef. Unlike barrier reefs, fringing reefs grow outward into deeper water right from the shore or coastline. Fringing reefs are the coral reefs we find most often in the tropical ocean.

- [x] Air dry modeling clay
- [ ] Toothpick
- [ ] Plastic tablecloth
- [ ] Yellow or brown acrylic paint
- [ ] Paper plate
- [ ] Paintbrush

**Weekly materials list**

Name: ______________________________

The last type of coral reef is called an atoll (said this way: ăt-åwl). Atolls are very interesting because these reefs have a circular or oval shape. Atolls begin as fringing reefs that form around an inactive volcano. Inactive volcanoes don't erupt anymore. Eventually, these old volcanoes sink beneath the surface of the ocean. As the coral reef continues to grow, a round reef and island form around the place of the old volcano. Many atolls have a beautiful lagoon or small island at the center.

Let's see if we can identify each type of coral reef!

What type of reef do you see in each image? Write **barrier reef**, **atoll**, or **fringing reef** on the lines to label each reef.

1. 

______________________________

______________________________

2. 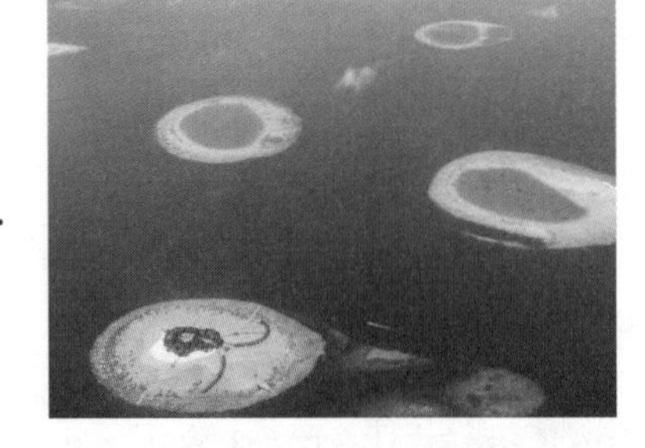

______________________________

______________________________

3. 

______________________________

______________________________

We're going to continue exploring coral reefs together today. Are you ready to get started? Coral is an important and beautiful part of the reef. I've been looking at pictures of coral, and there are so many different kinds! Some coral spreads out to create a flat surface, others grow tall pillars, while others have intricate patterns and look very delicate. And we haven't even started to talk about all the different colors! What is coral anyway, Hannah?

**materials needed**

- Air dry modeling clay
- Toothpick
- Plastic tablecloth
- Yellow or brown acrylic paint
- Paper plate
- Paintbrush

Great question. We can look at taxonomy first to see if we can find the answer. Which kingdom do you think coral is classified in, friend? Here is the answer: corals are classified in the animal kingdom.

Really? The pictures I saw of corals certainly don't look like any animals I've seen before!

True — but let's go deeper. The pictures of corals that you saw are actually a colony, or community, of tiny creatures called polyps (said this way: pŏl-ĭps). Polyps cannot produce their own food like plants, which makes them part of the animal kingdom.

Polyps are classified with other marine animals such as sea anemones and jellyfish in the Cnidaria (said this way: nī-dair-ē-ŭh) phylum. Polyps have a body that looks kind of like a tube with tentacles at the top. They use their tentacles to sting and stun their prey, like zooplankton. Once they have captured prey with their tentacles, they eat it through their small mouth.

While polyps could live on their own, most prefer to create colonies with other polyps from the same species. Together, hundreds or thousands of polyps living together in a colony create the hard coral structures we see in a coral reef. As the polyps attach themselves to form a colony, they use calcium from the ocean water to create a hard, calcium carbonate skeleton. This skeleton allows the polyps to anchor and protect themselves as they continue to grow.

Calcium carbonate has another name — it's also known as limestone!

Right! Eventually, the older polyps die — but the limestone skeletons they created remain in place. New polyps will attach and grow on the limestone coral structures. This is how the coral reef continues to grow and expand. Because of their hard skeletons, these corals are called hard or stony corals. A few species of hard coral are brain coral, pillar coral, and staghorn.

Hey, let's create our own model of brain coral today!

sun polyps

## Activity directions:

1. Spread out a plastic tablecloth to protect the surface you're working on.
2. Roll out a ball shape about 2–3 inches across. Flatten the bottom of the ball so that it stays in place.
3. Place the ball on the paper plate.
4. Look at the image of brain coral above and observe the intricate line design. Doesn't it look like a brain? Use the toothpick to create the line design on your model brain coral.

brain coral

5. Once you're done designing, set the paper plate in a safe place and allow your model to dry. Depending on how thick the clay is, it may take 1–3 days to dry.

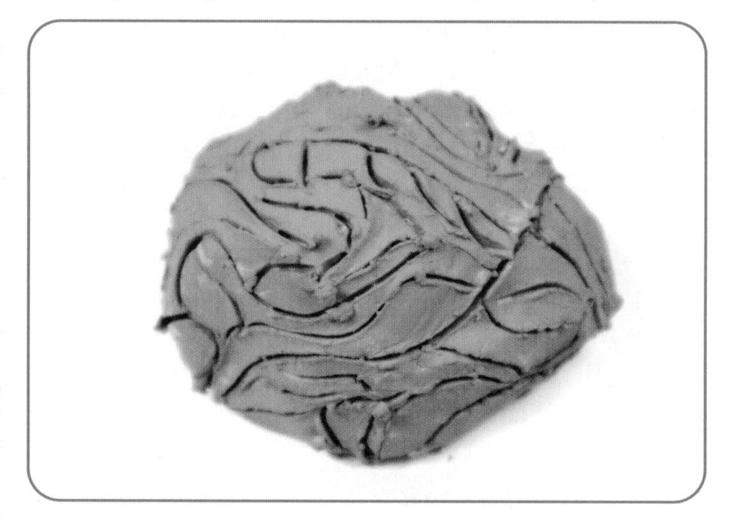

6. Once the model is dry, spread out the plastic tablecloth to protect the table. Paint your model yellow or brown and allow it to dry. Then place your brain coral in a safe place and keep it for next week's project.

Welcome back for another science adventure, friend! I've enjoyed learning about polyps and coral structures this week, haven't you? As we've been learning, I've noticed that coral reefs are full of such beautiful colors. I'm wondering now, what makes the coral colorful?

Ah, I'm glad you asked, Ben! I mentioned at the beginning of this week that God designed relationships between many different things that live in an ecosystem. One of the relationships we see often in creation is called a **symbiotic** (said this way: sĭm-bē-ŏt-ĭk) relationship. In a symbiotic relationship, two organisms live together in a close relationship.

The answer to your question, Ben, lies in a symbiotic relationship. You see, polyps actually don't have a color of their own. Instead, they receive their color from a relationship between the polyp and a type of algae called zooxanthellae (said this way: zōh-ŭh-zăn-thĕl-ŭh).

Wait, so it's the algae that gives coral its color?

Yep! Zooxanthellae algae live right inside the polyp. Because the polyp does not have a color of its own, it is the color of the zooxanthellae that we see in a coral reef.

Well, that is neat! But now I'm wondering, why would the polyp allow the algae to live inside the tissues of its body?

Good question. Both the polyp and the zooxanthellae receive a benefit from each other. The zooxanthellae use sunlight to photosynthesize, and the polyp will receive some of the sugar that is produced through photosynthesis. This provides nutrients for the coral. In return, the polyp gives the zooxanthellae a safe, protected place to live.

This type of symbiotic relationship is called **mutualism** (said this way: myoo-choo-ŭh-lĭz-ŭhm). Mutualism is a symbiotic relationship in which two organisms receive a benefit from each other. Scientists have observed that this relationship is especially important to the coral. When large storms or other factors affect the conditions on the coral reef, it can place stress on the coral.

Stress is a word that means to strain or pressure. In creation, the weather, disease, or predators can place stress on living things and may cause them to become unhealthy.

Right. When coral is stressed, it can cause the coral to force out the zooxanthellae. When this happens, the coral turns white and becomes unhealthy. We call this coral bleaching, and it can even cause large portions of a coral reef to die. Marine biologists are still studying what causes coral bleaching and how people may be able to help restore the coral reef when this happens.

Wow, relationships really are important in God's creation. Thanks for sharing with us today, Hannah!

Name: ______

apply it

1. What is mutualism?

2. Can you think of any other relationships in creation where two living things receive a benefit from each other? Hint: Think about pollination. What creature pollinates flowers?

3. Why do you think God may have designed mutualistic relationships in creation?

I had fun learning about coral reefs and the symbiotic relationship between polyps and algae this week. I can hardly wait to dive into learning more about the Great Barrier Reef next week!

Me neither! In the meantime, though, I've been thinking about mutualism. Remember, mutualism is a symbiotic relationship in which two organisms receive a benefit from each other. Polyps and zooxanthellae both receive a benefit through their relationship. Can you think of any other mutualistic relationships in creation, Hannah?

Well, bees and flowers are one example. The bees pollinate the flowers, and they receive a tasty meal from the flowers in return. Sea anemones and clownfish also have a mutualistic relationship that we are going to learn about next week.

That will be fun! Learning about mutualism this week reminded me of the relationships that God has placed in my life. You're my older sister, Hannah, but you're also my friend. We get to learn from each other, help each other, and build one another up.

That is true! Our relationship helps both of us to learn and grow in our faith. It reminds me of what Solomon wrote in Ecclesiastes 4:9–10,

*Two are better than one, because they have a good return for their labor: If either of them falls down, one can help the other up. But pity anyone who falls and has no one to help them up.*

God designed relationships to be an important part of creation — and they are an important part of our lives too. Our relationships with our family and friends are one way we learn more and grow in our faith together. When we have questions or feel sad, our friends and family can also help us and remind us of what God's Word says.

That's why it's important to choose friends who build up our faith and to help our friends build up their faith as well. What a neat reminder from the coral reef. Thanks for sharing, Ben!

Do you have someone who helps you to grow in your faith? Be sure to thank them for their friendship! Look up Ecclesiastes 4:9–10 in your Bible. If you'd like, you can highlight these verses. Memorize Ecclesiastes 4:9–10 with your teacher or with a sibling.

I don't know about you, but I sure am ready to add a new page to our Science Notebook today!

Me too! This week we learned about the three types of coral reefs. Do you remember what they are?

Let me think . . . the first type is called a barrier reef, then there is the fringing reef, and the atoll.

Good memory! We also learned all about coral and the symbiotic relationship between polyps and zooxanthellae algae. We classify their relationship as one of mutualism. Hey, I have an idea — let's draw a picture of a coral reef in our Notebook today!

I like that idea. We can make it bright and colorful. Here is a sketch we can use as an example to create our own coral reef.

This is how each of our coral reefs turned out. Have fun creating yours, friend!

In your Notebook, write:

Coral polyps and zooxanthellae algae have a mutualistic relationship; they both receive a benefit.

Then draw a picture of a coral reef.

Learning about mutualism this week also reminded us that our relationships are important. Relationships are one way we can learn more and build up our faith — and help others to do the same. Copy Ecclesiastes 4:9–10 on the back of your Notebook page as a reminder.

*Two are better than one, because they have a good return for their labor: If either of them falls down, one can help the other up. But pity anyone who falls and has no one to help them up* (Ecclesiastes 4:9–10).

# The Great Barrier Reef

Day 1

week 27

I'm so excited that you're here; I thought you'd never come!

**Note**

The student will complete several projects during the course of this week. For a full list of required materials, please refer to the Master Materials List.

**materials needed**

- Shoebox
- Plastic tablecloth
- Acrylic paint
- Paintbrush
- Water

Hello, friend! Ben is a little excited to begin our science adventure for this week. We're going to learn about the largest barrier reef in the entire world — the Great Barrier Reef!

I'm not a little excited, I'm a lot excited! We have so much to explore — and we'll get to make our own model coral reef at the end of this week.

What are we waiting for, then? Let's get started. The Great Barrier Reef is found off the coast of eastern Australia in the Coral Sea. It is the largest barrier reef in the world and stretches for over 1,400 miles.

That is one big coral reef!

Actually, the Great Barrier Reef is a chain of over 2,500 different reefs and hundreds of islands. It is so large that it can be seen from outer space.

I guess that is why we call it great, huh?

It must be! The Great Barrier Reef is one of the world's most beautiful and diverse ecosystems. It is home to around 600 types of coral and thousands of different species of marine life like fish, sharks, whales, turtles, octopuses, manta rays, and so much more. We'll learn about a few of the marine creatures found around the Great Barrier Reef this week. But first, let's talk a little more about coral.

One of the most common types of coral found in the Great Barrier Reef is called staghorn.

Hmm, I've seen pictures of staghorn coral in my books about the ocean. This type of coral has branches that extend like tree branches and twigs — it looks kind of like a deciduous tree in the fall.

Right! Its shape has also been compared to the antlers on a deer. Though it looks like a tree, remember that staghorn coral is actually part of the animal kingdom. These corals are made from hundreds or thousands of polyps living together in a colony. Staghorn coral can grow to be up to five feet tall. It can also grow tightly together and create a dense staghorn thicket.

Staghorn provides shelter and a habitat for other living things in the Great Barrier Reef. This type of coral is often tan, light brown, or gray, though you may occasionally spot brighter colors.

I have an idea — let's get started on our coral reef model today! We can paint the inside of our shoebox and then paint some staghorn coral as well. Ready to get started?

## Activity directions:

1. Spread out a plastic tablecloth to protect your surface. Paint the inside of your shoebox blue then rinse out your paintbrush. Allow the inside of the box to dry.
2. Paint groups of staghorn coral on the inside of the shoebox. You can use tan, gray, or another color of paint to create your staghorn corals.
3. If you'd like, you can paint seaweed, seagrass, or more types of coral near the bottom of the shoebox model. Rinse out your brush and allow the paint to dry.

4. Set your shoebox in a safe place — we'll finish our coral reef later this week!

Welcome back, friend! Today, we're going to begin learning about some of the other marine animals that live in the Great Barrier Reef. Ben has been studying so that he can tell us about two special creatures. What are we going to learn about today, Ben?

**materials needed**

- Scissors
- Toilet paper roll
- Colored tissue paper
- Glue stick
- Plastic tablecloth

The sea anemone (said this way: ŭh-nĕm-ŭh-nē) and the clownfish. Let's start with sea anemones first. Sea anemones are invertebrate polyps. They are classified in the Cnidaria phylum alongside coral polyps and jellyfish.

There are over 1,000 different species of sea anemone found in oceans around the world — some species can even be found in tide pools! Anemones can be smaller than a quarter or grow to be over five feet wide. Like coral polyps, anemones have a tube-like body and tentacles. Most anemones anchor themselves in one place through a single foot called a pedal disc.

Neat! What do sea anemones eat?

These animals are carnivores; they'll eat zooplankton or other types of small fish. Sea anemones use their tentacles to sting passing prey. Their sting injects a poison called a neurotoxin (said this way: nyoor-ōh-tŏk-sĭn) into their prey. The neurotoxin paralyzes the prey so that it cannot escape. Once the prey is captured, the anemone's tentacles move it to its mouth.

Now, with all of those stinging tentacles and a poisonous neurotoxin, you might think that no other creature would want to live anywhere near it — but sea anemones actually have a symbiotic relationship with clownfish!

Wait, clownfish? Wouldn't the anemone just sting and eat the clownfish?

It can try, but clownfish are protected from the anemone's sting by a mucus coating. In other words, the clownfish can swim into and all around the sea anemone — but the stings don't affect it. This allows clownfish and anemones to create a mutualistic relationship. The anemone gives the clownfish a safe place to live, while the clownfish helps to keep the anemone clean and provides additional nutrients for it.

That is really interesting. Thanks for sharing with us, Ben!

Sure thing! But we're not done yet — I also have an idea for how we can create our own model sea anemone. We'll use our model later this week in our coral reef. Let's get started!

## Activity directions:

1. Spread out a plastic tablecloth to protect your surface. Cut the toilet paper roll in half.
2. Cut 2–3 sheets of tissue paper about 2–3 inches taller than the roll and 6–8 inches long.
3. Use the glue stick to spread glue on the bottom inch of the first sheet of tissue paper. Then, press the roll into the glue and roll the tissue paper around it.

4. Repeat this process to glue the second and third sheets of tissue paper to the roll.
5. Repeat the process to create the second sea anemone.
6. Stand the sea anemones up, with the roll sitting on the table. Use scissors to cut the top of the tissue paper in thin strips — the cut should go almost to the roll.
7. Fluff the tissue paper strips so that they resemble a sea anemone's tentacles.

8. Show someone your sea anemones and tell them what you learned about them. Then set your anemones in a safe place; you'll use them later this week!

Today we're going to learn about one of my favorite marine creatures, sea turtles! There are seven different species of sea turtles found around the world.

The seven species of sea turtle are named the green sea turtle, leatherback, flatback, olive ridley, Kemp's ridley, hawksbill, and the loggerhead.

Each of those species, except for Kemp's ridley, lives in or around the Great Barrier Reef. The most common type of sea turtle found in the Great Barrier Reef is the green sea turtle, so let's learn about it today!

The green sea turtle is one of the largest sea turtles in the world. Their smooth shell can grow to be up to five feet tall, and they can weigh up to 700 pounds.

Ooh, I have a fun fact for us! A turtle's shell is also called a carapace (said this way: kair-ŭh-pās). Turtles on land are known to pull, or retract, their heads and flippers into their carapace to protect themselves. However, the green sea turtle doesn't have this ability.

Interesting! Now you may have noticed from looking at the picture of a green sea turtle that it doesn't really look green.

Hmm, that's right, now that you mention it. So why do we call them green sea turtles?

Their name comes from the green-tinted fat they have underneath their skin. Adult green sea turtles are primarily herbivores, eating green seagrasses and algae. Scientists believe that their diet may be responsible for the color of their fat.

One amazing skill green sea turtles have is their ability to migrate. **Migrate** (said this way: mī-grāt) means to travel from one place to another, often over a very large distance. Green sea turtles are known to migrate hundreds or even thousands of miles away from the beach where they hatched. But, when it is time for a green sea turtle to lay her eggs, she will often return to the very same beach where she hatched.

Whoa, and they do that without a map or without anyone to give them directions!

Exactly! Mom told me she wouldn't be able to navigate back to the place where she was born without the help of a map or a GPS. But God designed the green sea turtle with the ability to travel thousands of miles and still return to the place it started. Though scientists have studied their migration for years and years, they still aren't sure exactly how the turtle is able to accomplish this.

Well, we can be sure of one thing — since God is their Creator, He knows exactly how they do it!

Name: ______________________

apply it

1. What similarities do you see between the species of sea turtles?

2. What differences do you notice between the species of sea turtles?

3. How do you think the green sea turtle is able to migrate such long distances?

4. How might you test your hypothesis for turtle migration if you were a marine biologist?

It's been fun to learn about just a few of the marine creatures that live in and around the Great Barrier Reef this week! I wish we had time to learn about all of the hundreds of marine creatures we find there.

Me too! But there is at least one other creature that I wanted to talk about together. This creature belongs to the animal kingdom and is part of the Mollusca phylum. It has eight tentacle arms — can you guess what I'm talking about, Ben?

Hmm, an octopus?

You got it! It is called the mimic octopus, and it was discovered in 1998 near Indonesia. The mimic octopus can occasionally be spotted around the Great Barrier Reef.

While octopuses are known to be very intelligent and have the ability to change colors and patterns to camouflage into their surroundings, the mimic octopus goes a step further. This octopus can change its color, pattern, shape, and even its behavior to mimic other dangerous predators.

Mimic is a word that means to imitate or to act like something else. What creatures can the mimic octopus imitate, Hannah?

Well, this incredible octopus can mimic at least 15 other types of marine life, including the poisonous sea snake, sole, and the lionfish. When the mimic octopus is threatened, it quickly changes its shape, color, and behavior to scare away the threat or predator. But that's not all! The mimic octopus chooses to imitate the predator of the creature that is threatening it — how amazing is that!

Wow, God certainly gave this species of octopus an incredible amount of intelligence and talent! Its ability to mimic or imitate so well reminds me that we are to be imitators as well. But, unlike the mimic octopus that imitates many things, we are called to be imitators of God. Ephesians 5:1 tells us,

***Therefore be imitators of God as dear children*** (NKJV).

That is a great reminder, Ben! The mimic octopus can imitate and behave like many different things around it. It can be easy for us to imitate or behave like other people around us too — but this can get us into trouble! Rather than imitate everyone around us, the Bible tells us that we are to be imitators of Christ.

Exactly. As we learn more from the Bible and our faith grows, our lives also begin to look more and more like Jesus. This is called sanctification (said this way: săngk-tŭh-fŭh-kāy-shŭn). Sanctification is the process God uses to purify us from sin as we learn and grow in Him. As we are sanctified, the more and more our lives will imitate Jesus.

Talk to your family about ways God is sanctifying all of you. Look up Ephesians 5:1 in your Bible. If you'd like, you can highlight this verse. Memorize Ephesians 5:1 with your teacher or with a sibling.

Day 5

Ahoy, friend! Today is a special day — we're going to create our own model of the coral reef together.

There's only one thing I like better than adding a new page to my Science Notebook, and that is creating a model shoebox habitat!

Good, I'm glad. I have our art supplies, the crafts we've worked on, a few model marine creatures, and a shoebox right here and ready to go. Let's get started! Have fun creating your coral reef model, friend, and don't forget to share what you've learned with someone else!

**materials needed**

- Shoebox
- Sand
- Plastic tablecloth
- Craft model brain coral, anemones from previous lessons
- Small model marine life (turtle, manta ray, fish, etc.)
- Hot glue gun (optional)

## Activity directions:

1. Spread out a plastic tablecloth to protect your work surface.
2. Add sand to the bottom of the shoebox model.
3. Place your craft brain coral and sea anemones in the sand.

4. Add the small model marine creatures. Optional: If you'd like to add creatures to the upper part of your model, ask your teacher to help you hot glue them.

5. Share your coral reef model with your family. Be sure to tell them about the creatures you learned about, as well as the special designs God gave them.

**Bonus!** Take a picture of your coral reef model and ask your teacher to help you print it out. Then tape or glue the picture on the next page in your Science Notebook. Write **My Coral Reef Model** at the top of the page.

# Whales

## I

Ahoy, friend! Are you ready to join us on a new science adventure in the deep blue ocean? This week, we're going to learn about some of the biggest mammals in the ocean.

Wait just a minute, Hannah; we're studying marine biology. Don't you mean we're going to learn about the biggest fish in the ocean?

Nope, I definitely mean mammals. After all, fish aren't the only thing God created to live in the ocean. Did you know that water is also home to some of the largest mammals on the earth?

I didn't know that! Let's review what a mammal is before we get started. Mammals are living things that have hair on their bodies, feed their babies milk, need oxygen to breathe, and are vertebrates. Some examples of land mammals are cats, dogs, squirrels, horses, and cows.

Human beings are also classified as mammals in the field of taxonomy. Though humans share some features with animals, such as breathing air or having hair on our bodies, there is one feature we do not share with any other living thing. Do you remember what that is, Ben?

God made humans in His image — that is one feature that God only gave to people. Being made in God's image is what sets people uniquely apart from all other living things.

Alright, I can't wait any longer to find out what ocean mammals you're thinking of, Hannah!

Let's get started then — I'm thinking of whales! Whales are classified as mammals, and they are some of the largest animals on either land or in water. There are about 90 different species of whales found around the world, but they can be divided into two basic families: toothed whales and baleen whales.

We'll spend some time learning about baleen whales later on — let's start with toothed whales this week. As you might have guessed already, toothed whales have teeth. There are 76 different species of toothed whales around the world. A few species that you may recognize are the sperm whale, narwhal, and the beluga whale.

Stopwatch ✓

**Weekly materials list**

Name: ______________________

What about orcas? I know they have teeth!

Good question! Do you remember when we learned about taxonomy and classification? In taxonomy, whales and dolphins are classified together in the order Cetacea (said this way: sĕh-tŭh-sē-ŭh). This means that whales and dolphins are both grouped together as part of the large-toothed whale family.

However, we've also learned that taxonomy helps us to classify living things more specifically. Orcas are specifically classified as part of the smaller dolphin family. Orcas are the largest member of the dolphin family. Isn't it neat how taxonomy helps us to organize living things and group them together?

1. Have you ever seen a whale or dolphin in the ocean or at an aquarium?

2. What features can we look for to tell if a living thing is a mammal?

3. Scientists classify humans as mammals, but what makes human beings different from any other living thing?

4. Copy Genesis 1:27: "So God created mankind in his own image, in the image of God he created them; male and female he created them."

Hello there, we're going to learn a little about a toothed whale's anatomy together today.

I can help with that! **Anatomy** (said this way: ŭh-năt-ŭh-mē) means the parts of a body. Let's look at an orca's anatomy together. Remember, an orca is part of the dolphin family, and the dolphin family is grouped together with the larger toothed whale family. Though various species of whales and dolphins have different sizes and shapes, they all share a few basic parts.

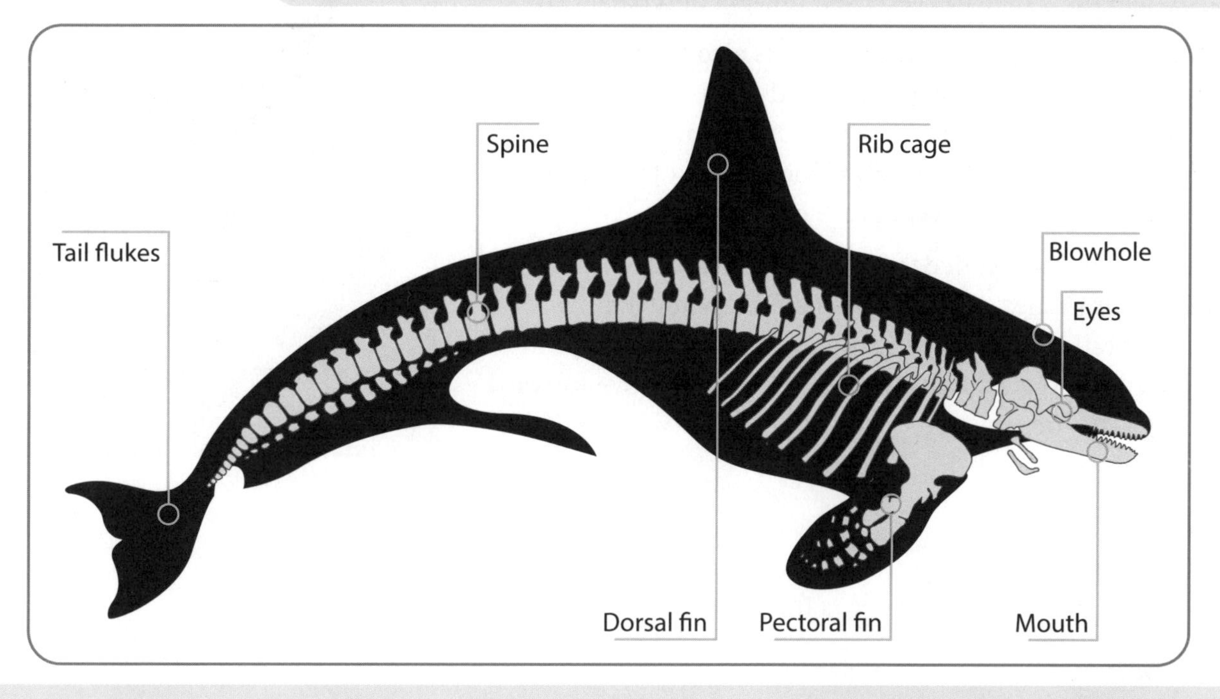


First, since whales are vertebrates, we know that they have a spine. The spine gives a whale or dolphin's body strength and stability, just like our spine does for us. Next is the ribcage. A ribcage has the important job of protecting a living thing's heart and lungs.

Can you feel your rib cage in your upper body, friend? God designed our ribcage to protect our heart and lungs as well. A human's rib cage is firm and sturdy. However, God designed the whale's ribcage to be firm enough to protect its heart and lungs but also flexible so that the whale can dive deep into the ocean.

Have you ever dived to the bottom of a swimming pool, friend? If you have, you may have noticed the pressure on your body from all the water above you. The deeper you dive into water, the greater this pressure becomes.

God designed a whale's body to be able to withstand the ocean's pressure as they dive deep. Without flexibility in their rib cage, the whale's rib cage could be crushed on a deep dive.

As mammals, whales and dolphins need to breathe oxygen from the air. This is where we find another one of God's amazing designs. At the top of a whale's head is its blowhole. The blowhole works like your nose does — it allows the whale to breathe in oxygen and exhale carbon dioxide at the ocean's surface as it swims along.

Name: ______________________

Now, let's learn the names for a whale or dolphin's fins. First, the fins on the side of its body are called the pectoral fins (said this way: pĕk-ter-ŭhl). The whale uses the pectoral fins to steer itself through the water. These fins also help the whale to keep itself upright and balanced.

Many species of whales and dolphins also have a triangular fin at the top of their body. This fin is called the dorsal fin. The dorsal fin helps it to swim quickly and efficiently through the water. Some species of whale have a tall dorsal fin, while others are quite short.

Finally, you may notice that the tail is made from two triangular shapes. We call each half of the tail a fluke. The full tail is referred to as flukes. Whales and dolphins move their powerful flukes up and down to move forward through the ocean.

1. What does the word "anatomy" mean?

2. How did God design a whale's rib cage different from a human's rib cage?

**materials needed**

- [ ] Stopwatch

Today, we're going to learn about sperm whales, and I'm so excited!

Well, let's get started then! The sperm whale is the biggest toothed whale in the ocean. They can grow to be over 50 feet long — that is longer than a school bus — and weigh an incredible 50,000–90,000 pounds. These amazing creatures live in oceans all around the world, from the warm tropical region all the way to the cold waters of the arctic and Antarctic.

No matter where it lives, though, it's easy to recognize a sperm whale by its big, rectangular-shaped head and long, narrow lower jaw.

That is true. You might even say that everything about a sperm whale is just plain big: it is the biggest toothed whale in the ocean, has a very large head compared to its body, and has the biggest brain of all the animals found on the earth. A sperm whale's brain can weigh up to 20 pounds!

Just imagine how quickly I could get my schoolwork done if my brain was that big.

Just imagine how big your head would be if your brain was that big! God designed sperm whales with the ability to dive incredibly deep into the ocean. In fact, sperm whales can dive around 2,000 feet down into the dark ocean waters. In order to dive that deep, they must also be able to hold their breath for 45–90 minutes.

Many creatures cannot dive deep into the ocean because the pressure of the water becomes too much and could crush them. Most people, for example, can only dive to around 40 feet — not even close to what the sperm whale can do!

Sperm whales dive so deep in order to hunt their prey. Sperm whales are marine carnivores that eat fish, squid, octopuses, and even some types of sharks. Sperm whales are known to hunt giant and colossal squid deep in the ocean.

Now, you might think that we would know a lot about these massive, toothed whales — but there is actually a lot scientists are still working to learn. Their ability to dive so deep into the dark ocean makes sperm whales really difficult to study and observe. Scientists still have a lot to learn about how God designed these whales to be able to dive so deep and about how they behave far below the ocean's surface.

Name: ______________________________

1. How long can you hold your breath? Use a stopwatch to find out — you can ask your teacher to help.

______________________________

2. How do you think a sperm whale is able to hold its breath for 45 minutes?

______________________________

______________________________

______________________________

3. What did you learn about sperm whales today that you think is the most interesting?

______________________________

______________________________

______________________________

4. If you were a marine biologist, what would you want to learn about sperm whales?

______________________________

______________________________

______________________________

Hello, friend. Did you have fun learning about toothed whales with us this week? I'm looking forward to learning more about whales as we continue our science adventures together. After all, we haven't even talked about baleen whales yet.

True, but I haven't forgotten about them! We'll be learning about baleen whales next week.

Good! In the meantime, there was something else I wanted to talk about. Dad and I watched a television show about whales last night. The marine biologists on the show used an underwater camera to capture amazing videos of different whales. I had fun watching the whales swim and play under the waves. The marine biologists shared a lot of observational science in the show.

Remember, observational science is science that we can see, experience, or observe, like how large a whale typically grows, where it lives, or what it eats.

Right! But then the biologists started to talk about where whales came from. They said that whales slowly changed, or evolved, from a land mammal about 50 million years ago.

Ah, it sounds like they were sharing an evolutionary worldview. Evolution is a theory about how the world and all that we see came to be. Evolution would say that one creature slowly changed into a different creature — like a land mammal changing into a whale — over millions of years.

Dad reminded me that the show had switched from observational science to historical science. Historical science cannot be observed, tested, or repeated. It is a theory that is affected by the scientists' worldview as they interpret the evidence we find. As Christians, though, our worldview is based on what the Bible says. In Genesis 1, we read that God created all marine and bird life on the fifth day of creation. In verse 21, it says,

*So God created the great creatures of the sea and every living thing with which the water teems and that moves about in it, according to their kinds, and every winged bird according to its kind. And God saw that it was good.*

According to the Bible, God created each kind, and He created them to reproduce after their own kind — and that is exactly what we observe in creation. No one has ever seen one animal change into a totally different kind of animal. However, we do find evidence that each kind of animal has stayed the same kind throughout history — just as God said. We'll be talking more about that soon!

As Dad and I talked about the show, I realized that understanding worldview is so important! As I read science books or watch shows on television, I can learn to recognize what worldview is being shared by comparing the information to what the Bible tells me. I can also learn to recognize observational and historical science by asking, "Is this information observable, testable, and repeatable?" If it isn't, it's most likely talking about historical science, which is based on a worldview and cannot be proven.

As you read about science in books or watch shows on television, practice asking yourself, "Is this information observable, testable, and repeatable?" Look up Genesis 1:21 in your Bible. If you'd like, you can highlight this verse. Memorize Genesis 1:21 with your teacher or with a sibling.

Hey there! I hope you brought your Science Notebook with you today because it's time to add a new page!

We learned about the sperm whale this week, and I think it's the perfect thing to draw in our Notebook.

I have a simple sketch right here that we can use for our example. We can also draw the ocean around our whales! Let's get started.

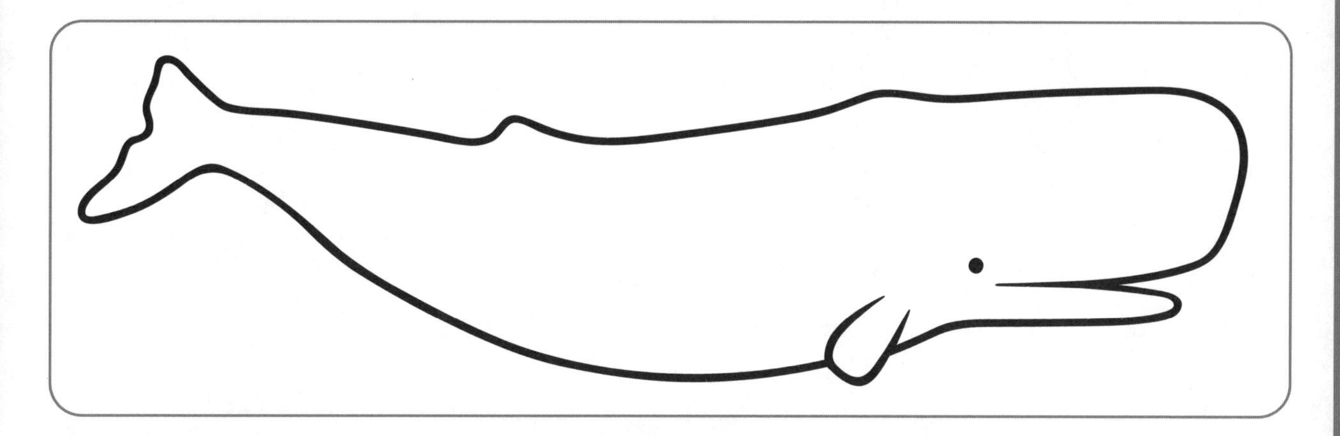

Here is how our drawings turned out. Remember to have lots of fun creating your drawing. We can't wait to see it!

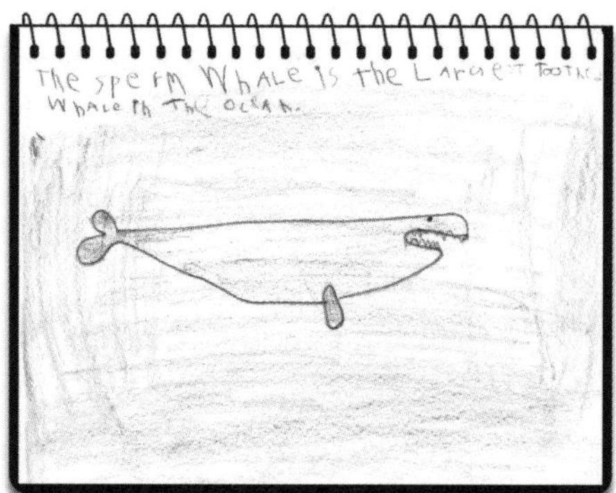


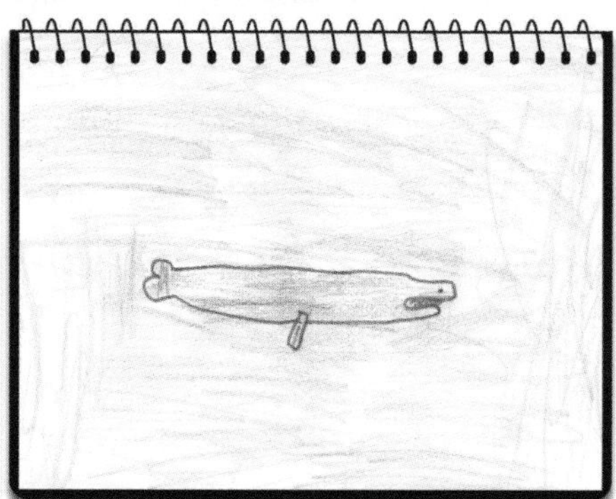

In your Notebook, write:

The sperm whale is the largest toothed whale in the ocean.

Then draw a picture of a sperm whale.

Learning about toothed whales this week also reminded us that as we study science, we can compare the information we learn to what the Bible tells us. The Bible is the foundation of our worldview. Copy Genesis 1:21 on the back of your Notebook page as a reminder that God created the whale kinds.

*So God created the great creatures of the sea and every living thing with which the water teems and that moves about in it, according to their kinds, and every winged bird according to its kind. And God saw that it was good* (Genesis 1:21).

week 29

# Whales

# Day 2

Welcome back, friend! Are you ready to jump back into our exploration of whales? We're almost ready to begin exploring baleen whales together. But first, there's one other amazing feature that God gave to toothed whales, and we're going to explore that today.

That's right! God gave toothed whales and dolphins the ability to use **echolocation** (said this way: ĕk-ōh-lōh-kāy-shŭhn) to help them hunt and navigate through the water. Echolocation is an incredible feature that allows bats, whales, and dolphins to use sound waves in order to "see" what is around them.

Ooh, I can't wait to learn more! How does echolocation work, Hannah?

Well, sound travels through air and water in a wave pattern. Toothed whales and dolphins create sound waves that we call clicks. These clicks travel out from the whale or dolphin's head and move quickly through the water — that is, until the sound waves bump into an object or obstacle. When the sound waves bump into something, they bounce back and return to the whale or dolphin.

Hmm, it's kind of like when I throw my rubber ball up against the tree in our yard. I aim the ball for the tree and then throw it. Once it hits the tree, it bounces back to me.

Exactly! Toothed whales have special areas of fat in their head that receive the sound waves as they bounce back. Then, just like our ears and brain work together to interpret the sounds that we hear, the whale's brain is able to interpret the sound wave echo that it receives back. It can use the echo to determine if there is prey, a predator, or an obstacle in the area.

Echolocation is an important feature for toothed whales and dolphins because it's not always easy to see in the ocean — especially for whales who can dive deep into the darkness! The whale's brain can interpret the sound wave echoes to help it "see" as it swims through dark or murky water.

Wow, what a wise design God gave toothed whales!

| Weekly materials list | |
|---|---|
| Blindfold | ✓ |
| Teacher or sibling | ☐ |
| Bowl or tray of water | ☐ |
| Clothespin | ☐ |
| Toothbrush | ☐ |
| Pepper | ☐ |
| Kitchen scale | ☐ |

I think so too! Sometimes, we as people are able to study and create designs or inventions that are similar to what God created first in nature. Sonar is used in ships and submarines to help them navigate through water. Sonar works in a similar way to echolocation, and it helps keep sailors safe.

Hmm, I have an idea for an echolocation game! Let's try navigating through a room without using our eyes — we'll need someone else to help us. Ready?

## Activity directions:

1. Ask your teacher or a sibling to help you with this activity.
2. Choose the room you will navigate through. Decide on your starting point and the finish line. Then set up a few obstacles to navigate around. You could use stuffed animals, pillows, or blankets.

3. Go to the starting point and put on the blindfold. Now call out, "Where should I go?" and listen for your teacher or sibling's reply.
4. Follow their directions then ask again, "Where should I go?" Use your teacher or sibling's directions to help you navigate through the room.
5. Once you've reached the finish line, you can trade places with your teacher or sibling if you'd like!
6. Once you are done playing, discuss the following questions with your teacher or sibling:
    - Was it easy or hard to navigate without using your eyes?
    - How did the directions help you navigate?
    - The directions you received back gave you information that helped you navigate — just like the echo a whale receives back helps it navigate. What do you think it would be like to be able to use echolocation like a whale?

### Note

You will need to call out directions for the student, such as, "Take three small steps forward," or "There is a pillow in front of you — take one big step to the right then two steps forward," to help the student navigate through the room.

Today is the day that we're finally going to jump into learning about baleen whales, right, Hannah? I'm so excited!

Right — I won't make you wait any longer.

Yay! I already know that there are 14 different species of baleen whales found in oceans all around the world. A few species of baleen whales are humpback, grey, right, and my personal favorite, the blue whale. Baleen whales are much like the toothed whales we studied last week except that God designed them with plates of baleen instead of teeth. Can you tell us what baleen is, Hannah?

**materials needed**

- Bowl or tray of water
- Clothespin
- Toothbrush
- Pepper

Sure! As I explain, it might help to think of baleen like the bristles on a toothbrush or a broom. Baleen is made of keratin (said this way: kĕh-rŭh-tĭn), which is what your fingernails and hair are made of as well. In baleen whales, God designed keratin to grow down in long plates from the top of the whale's mouth. These plates combine to form a row of baleen bristles. Depending on the species, the baleen plates may grow to be just a few inches long or over ten feet!

Wow! So, I know how teeth are used for eating — but how do baleen plates work?

Great question. Baleen plates are perfect for filtering food from the water — and that's exactly how baleen whales eat. Baleen whales eat small fish, plankton, and krill. Krill are invertebrates that look a bit like shrimp. Krill only grow to be around two inches long, which would make them quite difficult for toothed whales to hunt and eat. It isn't a problem for baleen whales, though, because baleen works as an excellent ocean filter. When a baleen whale locates a swarm of krill, it opens its mouth wide and swims through the swarm. Its mouth fills with the ocean water, as well as the krill.

But the whale doesn't need to swallow all of that water. How does it swallow just the krill, plankton, or small fish it's holding in its mouth?

The whale closes its mouth and then uses its tongue to push the water through the baleen plates. The plates are perfectly designed to allow the water to pass through — but to trap the whale's dinner right in its mouth. Once the water has been filtered through the baleen, the whale can then swallow the food it has caught. Let's do an activity to see how baleen works!

## Activity directions:

1. Sprinkle pepper into the bowl or tray of water. You're going to use the toothbrush and clothespin to gather as much of the pepper as you can — the toothbrush will work like baleen, while the clothespin will work like the mouth of a toothed whale.

ASK PARENT FOR HELP

2. Write your name and the date on your lab report on the next page. Next, write the question: Will the toothbrush or clothespin be able to gather more pepper?
3. Write your answer in the Hypothesis section of your lab report.
4. Open and close the clothespin to "bite" the specks of pepper. Do you think it is easy or hard to capture the pepper this way? Write your observation in the "Things I observed" section of your report.
5. Now use the bristles on the toothbrush to filter the pepper from the water. Drag the bristles through the water. How much pepper do you capture? Do you think it is easy or hard to capture the pepper this way? Write your observation in the "Things I observed" section of your report.
6. Finish your lab report and record the results of your experiment.

Name ______________________________ Date ______________

# Lab Report

## Question

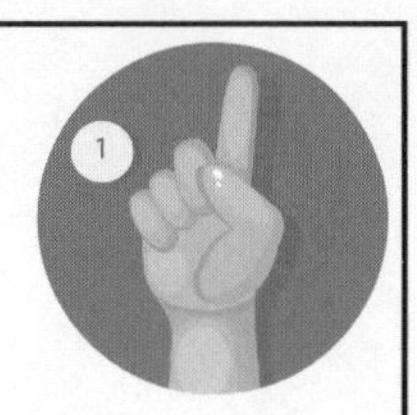


## Hypothesis

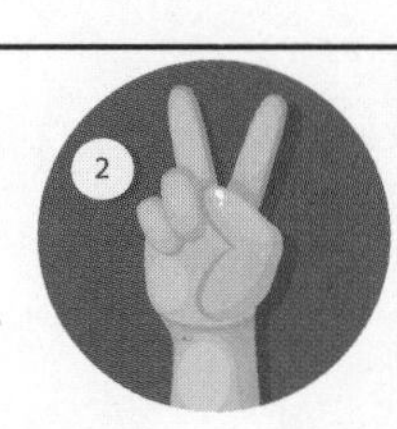


## Things I observed:

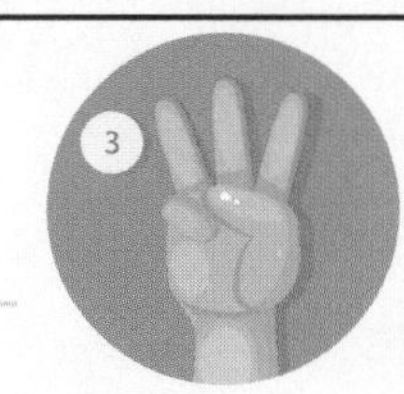


## Results

### What happened in the experiment?

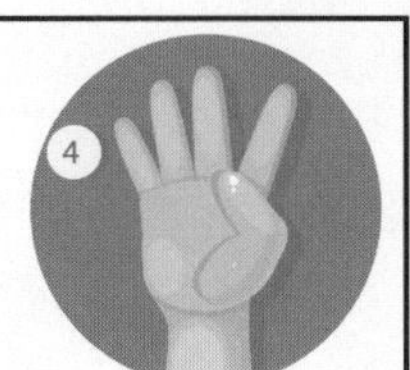


### Was my hypothesis correct?

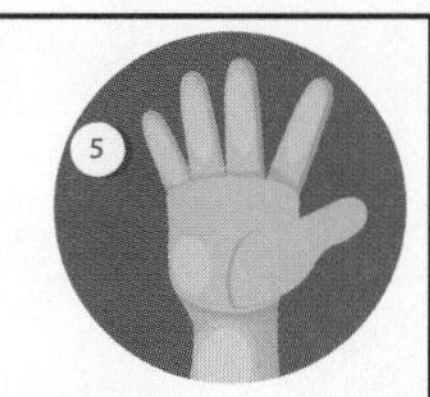

# Additional Lab Notes

Hello there, science adventurer! Today, we're going to talk about the largest animal on the earth: the blue whale.

Ooh, I can't wait! The blue whale is my favorite whale — I hope to be able to see a real one in the ocean someday. Blue whales can grow to be 80 to 100 feet long, which is longer than a semi-truck and trailer!

**materials needed**

- Kitchen scale

But that's not all. They can weigh over 300,000 pounds. That makes blue whales the largest animal in God's creation that scientists have ever been able to discover. Blue whales are even bigger than the dinosaurs were!

That is so neat! Something that we notice in the ocean is that marine life is often able to grow much bigger than the life we find on land. Why does that happen, Hannah?

Ah, the answer to that question lies in water. Have you ever watched someone or something float on the water? This is called buoyancy. **Buoyancy** (said this way: boy-ŭhn-sē) is the ability to float or be suspended by water. Buoyancy allows marine life to grow bigger because it helps support the creature's weight.

On the other hand, land animals aren't supported by buoyancy. That means they can only grow as big and heavy as their skeleton can support. If a land animal were to grow as big and heavy as the blue whale, for example, its weight and size would likely crush its bones and lungs!

And nobody likes a broken bone.

That's for sure. The ocean is also able to supply the large amount of food that big whales need to survive. For example, a blue whale may eat as much as 8,000 pounds of plankton, krill, and small fish in a single day.

Name: ____________________

Blue whales can be found in oceans all around the world, except for the Arctic Ocean. Like sea turtles, blue whales also migrate great distances in order to find enough food and locate safe places for their babies to be born.

In addition to being the largest animal in the world, the blue whale is one of the loudest animals in the world. Baleen whales communicate with each other through moans, groans, and songs that travel long distances through the ocean. The blue whale's song can be heard around 1,000 miles away! However, the human ear isn't designed to be able to hear most of the sounds a baleen whale makes without help from special equipment.

## Activity directions:

1. Ask your teacher to help you set up the kitchen scale to measure in pounds or ounces. Answer question 1 on the worksheet below.
2. Before you eat your next meal, weigh your meal on the scale. Answer questions 2–4 on the worksheet.

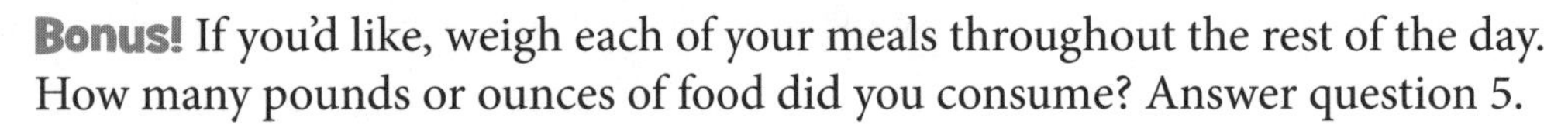

**Bonus!** If you'd like, weigh each of your meals throughout the rest of the day. How many pounds or ounces of food did you consume? Answer question 5.

1. A blue whale can eat 8,000 pounds of food in a single day! How many pounds of food do you think you eat in a day?

____________________

2. How much did your meal weigh? ____________________

3. Did it weigh more or less than you thought?

____________________

4. What do you think it would be like to eat 8,000 pounds of food in a day?

____________________

____________________

____________________

**Bonus!** Weigh each of your meals and snacks today. How many pounds or ounces of food did you eat all together?

____________________

____________________

____________________

I'm so glad we decided to explore marine biology together. It has been fun to learn more about marine life and the way God designed these incredible creatures!

I've enjoyed learning about whales during our last few adventures together. Last night, Ben and I were talking with our mom about whales — she was able to see a real blue whale in the ocean once!

That's why I hope to get to see one someday too. Mom told us about how whales leave "footprints" on the surface of the ocean. Hannah and I thought that was crazy; how can a whale leave a footprint in water?

But then she explained that after the whale comes to the surface to breathe, they'll dive back down into the ocean. When they dive, their powerful flukes form a wave underwater that rises to the surface and disrupts the water's wave pattern. This disruption leaves a circular- or oval-shaped area of the water where the whale was. This area looks different than the rest of the water. This is a picture our mom took of a blue whale's footprint!

It kind of looks like oil was spilled on the water there.

It does! Eventually, the water will return to its normal wave pattern. But in the meantime, these whale footprints show where the whale has been and can help scientists and whale spotters locate them.

Just as the whale leaves an impact on the water around it, our lives can also have an impact on others. We call the impact our lives have on others our influence — and we can have a good or a bad influence on others.

A good influence encourages others to follow Jesus, obey the Bible, grow in their faith, and make wise decisions. On the other hand, a bad influence can draw others to disobey the Bible's instructions, stop following Jesus, and make unwise decisions.

I want to be a good influence for my friends and family — to build them up in their faith! It reminds me of what Proverbs 27:17 says:

*As iron sharpens iron, so one person sharpens another.*

Another way to say that is that two people can build each other up and become stronger together.

And just like we can be a good influence on others, it's important to remember that others can also have an influence on us. This is why it is important to choose friends who will help us to continue growing in our faith. How do you think you can help others grow in their faith, friend? Be sure to talk about it with your teacher or with a sibling!

Look up Proverbs 27:17 in your Bible. If you'd like, you can highlight this verse. Memorize Proverbs 27:17 with your teacher or with a sibling.

Today is the day — let's add a new page to our Science Notebook!

Are you sure, Ben? I don't know if you really want to add a new page.

Hannah, don't be silly — this is my favorite day! We learned about echolocation and baleen whales this week. I really want to draw a picture of a blue whale in our Notebook.

Okay, that sounds like fun!

Here is a sketch of a blue whale that we can use for our example. You can also add the ocean around the whale and draw some fish if you'd like.

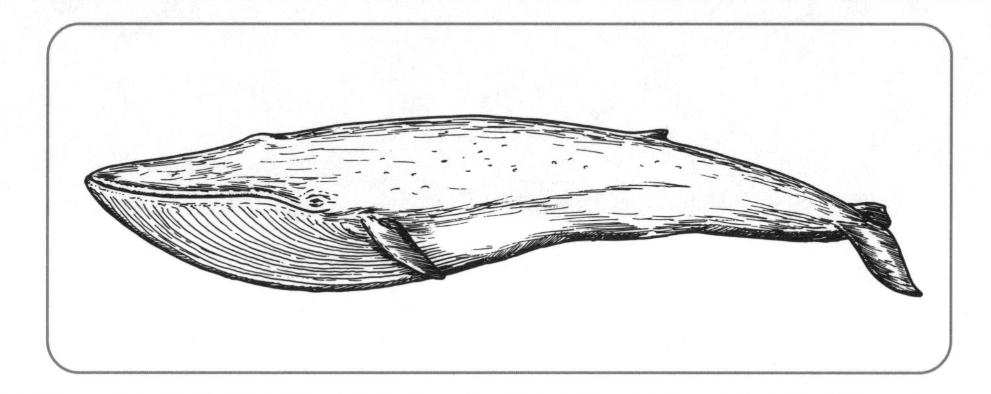

Would you like to see how our drawings turned out? Here they are! Hannah says that my blue whale looks very happy. Our little brother Sam drew a baby blue whale. Have fun with your drawing, friend!

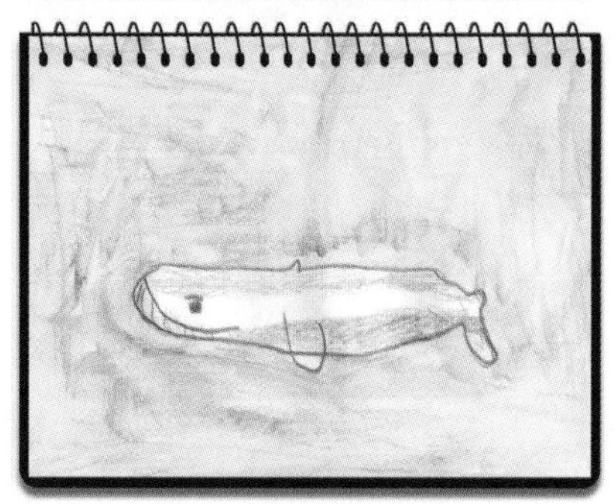

In your Notebook, write:

The blue whale is a baleen whale.

Then draw a picture of a blue whale.

Learning about whale footprints this week also reminded us that our lives can have an impact on others. Copy Proverbs 27:17 on the back of your Notebook page as a reminder.

*As iron sharpens iron, so one person sharpens another* (Proverbs 27:17).

week 30

# Conservation

Welcome back, friend! We've had fun learning about whales in our last few adventures together. Did you know that some species of whale, like the sperm whale and the blue whale, were nearly hunted to extinction (said this way: ĭk-sting-shŭhn) in the past? **Extinct** (said this way: ĭk-stingkt) means that a species of living thing is no longer found alive anywhere on the earth.

Oh, like the dinosaurs or the woolly mammoths. It is a sad thing when a creature becomes extinct because we lose a part of God's creation.

This is one reason why it is important to wisely care for God's creation. The work of caring for, protecting, and preserving God's creation is called **conservation** (said this way: kŏn-ser-vāy-shŭn). The word conserve means to protect or keep from wasting something.

Conservation is one of the important jobs marine biologists do. But before we start talking about the ways marine biologists help to conserve creation, let's talk about why this is important work. Where should we look first to find our answer?

The Bible, of course! Let's read the instructions God gave to Adam and Eve in Genesis 1:28,

*God blessed them and said to them, "Be fruitful and increase in number; fill the earth and subdue it. Rule over the fish in the sea and the birds in the sky and over every living creature that moves on the ground."*

So, God told them to rule over the living creatures. Another way to say this is that God gave human beings dominion over creation. Dominion means the power or authority to rule over something.

However, it's important to know that dominion doesn't mean that we get to do whatever we want with God's creation. Instead, it means that we are to care for it the same way that God cares for us. We want to tend to God's creation and manage it wisely — to be good stewards.

Name: ______________________________

The word steward means to manage, look after, and care for something that belongs to someone else. When the foundation of our worldview is built on the Bible, we recognize that God created the earth and all that is in it. Creation belongs to God — but He has given it to us to manage, to steward.

One job that marine biologists do is helping to conserve and care for the creatures God created to live in the ocean. I may not be able to do that job, but there are other ways I can be a good steward of God's creation even as a kid.

That's because being a good steward starts with caring for the things God has placed in our care right now, like our pets, toys, home, and neighborhood. We'll be talking more about conservation and stewardship together this week, but for now, let's think about what God has placed in our care.

apply it

1. What does conservation mean?

______________________________

______________________________

______________________________

______________________________

2. What is something God has placed in your care?

______________________________

______________________________

______________________________

______________________________

3. How do you care for it as a wise steward?

______________________________

______________________________

______________________________

______________________________

Hello, friend! Let's dive right into our adventure today. Last time, Ben mentioned that sperm whales were nearly hunted to extinction. Can you tell us more about what happened, Ben?

Yes! Before we begin, though, it's important to remember what we learned in Genesis 9:3 a few weeks ago.

Oh, I remember! That verse says that God also gave the animals as food for people in our fallen world.

Right. We receive food and many other useful things from plants and animals. As stewards, it is important that we manage creation wisely. At times, people have taken or harvested too much of a particular plant or animal. When this happens, it becomes an endangered species. This means that there aren't very many of that species of plant or animal left in the wild — it could easily go extinct. This is what happened to the sperm whale during the 17th and 18th centuries.

The sperm whale gets its name from a special substance called spermaceti (said this way: spur-mŭh-sĕt-ē). Spermaceti is found in the whale's head and is a very interesting substance. Though scientists have studied sperm whales for a long time, they still aren't exactly sure what the purpose of spermaceti is. Two theories are that it is useful for echolocation or helps the whale to be able to dive so deep.

Whatever purpose God created spermaceti for, people certainly found it very valuable because it can be used to make things like candles or used as oil for lamps. This made spermaceti very important for life in the 17th and 18th centuries. During this time, sperm whales were hunted in great numbers so that the spermaceti could be harvested. Eventually, so many sperm whales had been taken from the ocean that they nearly went extinct.

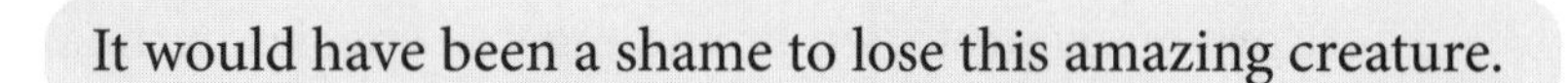

It would have been a shame to lose this amazing creature.

Definitely. Once people realized that the sperm whale was endangered, they knew that things needed to change — it was time to conserve the whales. So, the sperm whale became a protected species. When a species is protected, it means that it is no longer allowed to be hunted or that it can only be hunted under certain conditions.

Scientists believe that at one time, there were over one million sperm whales in the ocean! Today, scientists estimate that there are around 300,000. Though their numbers are increasing, the sperm whale is still a protected and endangered species today.

Name:

Thanks for sharing with us, Ben! This is a good example of how conservation works. Marine biologists study life in the ocean, as well as how each species interacts with one another. Marine biologists can track and estimate the number of a species. This information can help us discover when a species needs help.

apply it

1. Look up Genesis 9:3 in your Bible then copy it here:

2. What does it mean when a species is protected?

3. There are many protected or endangered species in the world today. Can you think of any other animals that are protected or endangered? Hint: Think of some of the big animals you may see at the zoo.

Day •••

As we've been learning this week, I've been thinking about stewardship. We learned earlier that God gave us dominion to rule over and be good stewards of the earth. I'm just a kid, though! How can I be a good steward of God's creation?

Great question, Ben. We may not be able to do the work that a marine biologist is able to do, but there are many other ways we can be good stewards of creation and help to conserve it. Let's talk about a few of those ways today.

The first thing we can do is continue to learn all that we can about God's amazing creation. When we understand how God designed His creation to work, we can use that knowledge to help us better care for the plants and animals around us.

Hmm, like when we brought our cat, Bell, home! Mom gave me a book to help me learn more about cats. I learned what cats should and shouldn't eat, as well as how to care for her. I also learned how to tell when Bell isn't feeling well so that I can help her.

Mom also teaches us about the plants in her garden and in the yard around us so that we can care for them. Learning is an important way we can become good stewards. The second thing we can do is to take good care of the things God has given us.

We talked a little about that earlier this week. I can take good care of my belongings and room, as well as our pets. I can make sure they have plenty of food and water, clean up after them, and give them plenty of love.

Exactly. Another thing we can do is conserve resources. God provided many resources in creation through plants, animals, water, land, trees, energy, and more. We receive many benefits from each of these, and we can use them wisely to benefit creation.

Water is one of the resources we have on earth — and we all need it! Keeping water clean to drink and play in is important. One way we can do that is by making sure our garbage is taken care of. When there is garbage where it doesn't belong, we call it litter. Litter can pollute or contaminate land and water in an area. When this happens, it can cause harm to the people, plants, animals, birds, and insects that live there.

And don't forget we can also work to not waste resources like food, water, and energy! These are all good ways we can be good stewards of God's creation right now — and learn how to continue being good stewards of God's creation as we get older.

Name: ______________________

apply it

1. What is your favorite thing to learn about in God's creation?

2. What would you like to learn more about in God's creation?

3. What is one way you can be a good steward right now?

I'm glad we've been able to learn about conservation this week, and I'm looking forward to talking more about it next week! In the meantime, I've been thinking about sperm whales. We learned that they were nearly hunted to extinction in the 17th and 18th centuries. But when people realized the mistake they were making, they changed their minds and began to protect the whales instead. This reminded me of Acts 3:19,

*Repent, then, and turn to God, so that your sins may be wiped out, that times of refreshing may come from the Lord.*

Huh. Why does the story I told about sperm whales remind you of that verse, Hannah?

Well, in the Bible, God tells us what His standards are and how we are to live. Sin means that we have missed the mark — that we've failed to live by God's standards. The Bible also tells us that when we have sinned against God, we are to repent.

Repent means to change your mind because you regret something you said, did, or thought. It means that we agree with God that what we did was wrong. It also means that we want to change what we are doing to follow His standard.

In the story you told us, Ben, people realized they had made a mistake by hunting so many sperm whales. They could have chosen to do nothing — if they had, we may not have any sperm whales in the ocean today. But instead, people chose to change their minds on what should be done with sperm whales. They decided to protect them instead of hunting them. They changed their minds and their behavior.

In the same way, when we sin against God, we can realize that we've made a mistake and need to repent. Repentance is more than simply saying "I'm sorry" — it is asking God to help us change so that we can follow His standard.

Ah, now I understand! And when we repent, the Bible tells us that we are forgiven for our sins through Jesus. 1 John 1:9 says,

*If we confess our sins, he is faithful and just and will forgive us our sins and purify us from all unrighteousness.*

I'm so very grateful that God shows us mercy and forgives our sin when we repent!

Look up Acts 3:19 in your Bible. If you'd like, you can highlight this verse in your Bible. Memorize Acts 3:19 with your teacher or with a sibling.

Well, Ben, do you know what day it is?

Don't be silly, Hannah — of course I know what day it is! It's time to add a new page to our Science Notebook.

This week we started to explore conservation. We learned that God gave us dominion over creation and that we are to be good stewards of what He has placed in our care.

I was thinking we could draw a picture of something God has given us to care for. It could be a pet, plant, favorite toy, our room, or even the yard!

I like that idea. Sam will want to draw a picture of the fish tank. I think I will draw a picture of our dog, Sadie.

Ooh, and I'll draw a picture of Bell. Let's get started on our drawings. What will you draw, friend? We can't wait to see it!

In your Notebook, write:

I can be a wise steward of what God has given me.

Then draw a picture of something God has given you to care for.

Learning about how sperm whales became protected reminded us of how repentance works. Copy Acts 3:19 on the back of your Notebook page as a reminder.

*Repent, then, and turn to God, so that your sins may be wiped out, that times of refreshing may come from the Lord* (Acts 3:19).

week 31

# Scientific Method in Marine Biology

Day 1

Hello there, friend! Can you believe we've reached our 31st science adventure together already? Time sure flies when you're having fun! In our adventures so far, we've talked about the history of science and the scientific method, learned about lab reports, measurements, matter, the periodic table of elements, and molecules.

And don't forget about the carbon cycle, mixtures, worldview, living things, classification, and now marine biology! Exploring marine biology together has made me curious — how are marine biologists able to use the scientific method to learn more about the ocean? It seems like the scientific method is easy to use in a field of science like chemistry, but how can it be used in marine biology?

Good question, Ben! Let's talk about that this week. First, we'll need to review the scientific method.

The scientific method helps us study science in an organized way as we ask questions, test our ideas, and develop conclusions. We can review the five steps that we use in the scientific method in the image on the right.

Make an Observation

Ask Questions

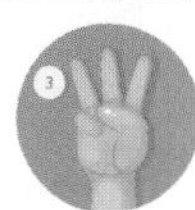

Create a Hypothesis

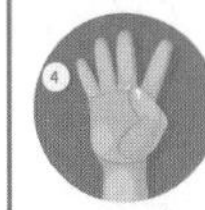

Test it (Experiment)

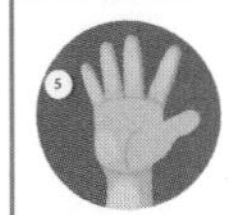

Share the Results

Exactly. Now let's get back to marine biology. Over 70% of the earth is covered by ocean — just think of how many hundreds and thousands of living things there are in the ocean! Marine biologists have an incredible amount of water and marine life to learn about as they work to classify marine life, understand its behavior, monitor and explore the ocean, and even solve ocean mysteries.

Now, can you imagine trying to do all of this work without an organized process to follow, like the scientific method?

I think it would make their job much harder to do!

For sure. One important job marine biologists have is making observations and collecting data. Remember, observe means to see or notice something. A marine biologist who studies sea turtles, for example, may carefully observe their behavior and collect information about their size, habitat, diet, and travel patterns. This data can then be shared with other marine biologists who may ask more questions about sea turtles.

Name: ______________________________

And then they can create a hypothesis and begin testing it!

Yup! Hey, this gives me an idea — let's practice making some of our own observations today.

## Activity directions:

1. Choose a living thing to observe. You may choose a pet, birds in your yard, a plant, fish, or even a parent or sibling.
2. Answer question 1 on the worksheet below.
3. Carefully observe the living thing you've chosen for 5–10 minutes. What do you notice about its size, shape, color, sound, and behavior?
4. Write your observations in question 2 on the worksheet. Then draw a picture of what you observed.

ASK PARENT FOR HELP

apply it

1. I chose to observe:

______________________________

2. This is what I observed:

______________________________

______________________________

______________________________

______________________________

3. Draw a picture of the living thing you observed.

Hey, friend! Did you have fun making observations last time? Hannah observed our cat, Bell, and I observed the fish in my fish tank! I have neon tetras in my fish tank. They are my favorite kind of fish. I love their bright blue and red coloring.

What observations did you make about your fish, Ben?

Well, I noticed that they are about an inch long. They seem to like to swim and stay together in a group. I also watched carefully as they used their gills to breathe under water. But then I noticed something odd! Two of my neon tetras had strange white patches on them. I carefully observed the other fish, and they didn't have those patches.

Hmm, that sounds strange.

I thought so too, so I asked Mom to come look at the fish. She said they had an infection caused by bacteria or fungus.

Oh no!

Don't worry — Mom had medicine in the cabinet for the fish. She measured the right amount and added it to the fish tank. The infected fish should be better in a few days.

Wow! I'm glad you spent some time observing your fish, Ben.

Me too! The time I spent observing them led to these questions: "Why do two of the fish have white patches? What is causing them?" Someone must have had those same questions in the past and studied what caused the white patches. They created a hypothesis to answer their question, tested it, and shared the results. That's how we knew that the fish had an infection and what medicine would help them.

Your observations gave us one small example of how the scientific method can be used in marine biology, Ben. Marine biologists can study the chemistry of the ocean, waves and weather patterns, ocean geography, and marine life. Since the ocean is so vast, marine biologists often choose something to focus on. For example, the Sarasota Dolphin Research Program studies bottlenose dolphins around Sarasota Bay in the state of Florida.

Hmm, I read about them in a book! They've been observing and studying dolphins there since 1970. That's over 50 years of observing, asking questions, testing hypotheses, and sharing the results.

Yes, their research has helped people learn more about the behavior, health, communication, and travel patterns of bottlenose dolphins. The knowledge they've gained from their research has even been used to help dolphins in other areas of the world.

Name: ____________________

And do you know what is amazing? Even after so many years of observing and learning, they are still observing and learning more about bottlenose dolphins. God's creation is amazing — there is so much for us to learn about all of the different things He created!

Imagine you have the opportunity to observe bottlenose dolphins for a day. Answer the questions below.

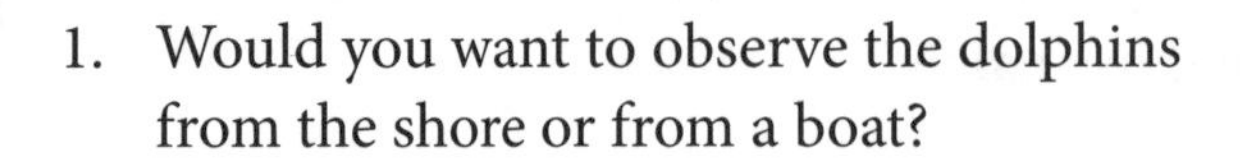

1. Would you want to observe the dolphins from the shore or from a boat?

2. Why?

3. What question would you have about bottlenose dolphins? For example, "How long are bottlenose dolphins?"

4. How do you think you could answer your question?

5. What do you think it would be like to study bottlenose dolphins for over 50 years?

We've been talking about the scientific method in marine biology this week. Let's dive deeper into conservation today.

I knew I should have brought my scuba diving gear along.

Oh, Ben, don't be silly. Marine biologists study life in the ocean, as well as how each species interacts with each other.

Hmm, that sounds like the field of ecology! **Ecology** (said this way: ĭh-cåll-ō-jē) is the study of the environment plants and creatures live in, as well as the relationships between living and nonliving things.

Marine biologists use many different tools to study marine ecology. For example, they may use a boat in order to observe marine life and their behavior at the surface. A marine biologist could also use a snorkel, mask, and fins in order to take a look under the surface of the water. Or they could use scuba diving gear in order to dive much deeper under water.

Did you know that the word scuba is an abbreviation? An abbreviation is a shorter way of writing something. Scuba stands for **s**elf-**c**ontained **u**nderwater **b**reathing **a**pparatus.

Neat! Divers often use underwater cameras and microphones to record pictures, videos, or sounds of marine life to share with others. Submarines are also used to travel deep into the ocean.

GPS trackers are another interesting tool. These can be attached to marine life — like a shark, whale, or dolphin. Once the tracker has been attached to the creature, it sends a signal back to the scientists that shows where the creature is. Scientists can observe and analyze these signals to find patterns. For example, if many whales travel to the same area around the same time, this could help a marine biologist learn about their migration pattern.

And don't forget about observations! A marine biologist can observe how many creatures are in an area and use this information to estimate how many creatures there are in the ocean.

An estimate isn't the exact amount of something, but it gives us a good idea of how many there are. This information can help us discover when a species needs help, like the sperm and blue whales. Or an estimate may reveal that there are too many of a particular plant or animal and it is causing harm — like the purple sea urchins in a kelp forest. When that happens, we can help to remove it.

All of these tools, the scientific method, and a scientist's observations help them learn what it looks like when marine life is healthy and things are in balance in the ocean.

Name: ____________________

Scuba divers have several tools to help them swim, breathe, and keep them safe under water. Find the name of each item in the word search.

| Wetsuit | Fins | Regulator | Tank |
|---|---|---|---|
| 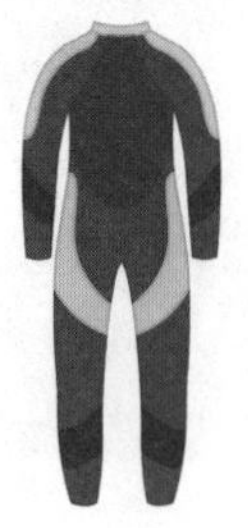 | 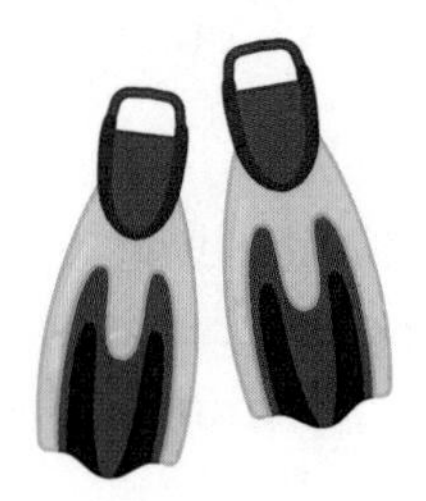 |  |  |
| Helps keep the diver warm. | Help a diver swim easily through the water. | Delivers air at the right pressure for the diver to breathe. | Stores oxygen for the diver to breathe. |
| **Gauges** | **Mask** | **Light** | |
|  |  |  | |
| Give the diver information about how much air they have left, how deep they are, or what direction they are traveling. | Allows the diver to see under water. | Helps the diver see through dark or murky water. | |

| | | | | | | | | | | | |
|---|---|---|---|---|---|---|---|---|---|---|---|
| W | Q | P | W | T | L | K | M | A | S | K | X |
| R | E | G | U | L | A | T | O | R | Z | V | G |
| J | A | T | N | Z | Q | L | W | Y | F | L | A |
| D | F | A | S | J | K | I | F | I | N | S | U |
| M | Z | N | T | U | E | G | T | O | R | Z | G |
| N | R | K | D | O | I | H | S | T | I | J | E |
| S | N | V | E | R | T | T | B | C | A | Z | S |

Why, hello there, friend! I've been thinking about the Sarasota Dolphin Research Program lately and the work that they do studying bottlenose dolphins. They observe many different dolphins each year, but they can recognize individual dolphins in the water by the shape of their dorsal fin. They've even given nicknames to some of the dolphins that they see quite often! Scrappy is the name of one dolphin they've often observed.

Scrappy was born in 1998 and has been spotted and observed by the Sarasota Dolphin Research Program hundreds of times. But once in 2006, they noticed that something wasn't quite right. Scrappy had gotten his body tangled in a discarded swimsuit. The scientists hoped that it might fall off or that he would be able to free himself from it eventually — but after a few weeks, it was clear that the swimsuit was not going to go anywhere. Worse than that, it was digging into his skin and causing a wound.

Oh no!

Well, the researchers got to work. They received special permission to capture Scrappy safely so that the swimsuit could be removed. Once it was removed, he was released back into the ocean. His wound healed right up, and they've spotted him many times since.

Wow, it must have been exciting to be able to touch and help him!

I think so! It's one way these scientists have helped to conserve God's creation. But our adventure this week also reminded me that being a good steward isn't always a lot of fun. Sometimes it is a lot of work — or even really smelly.

Like when you have to clean out our cat's litter box?

Exactly! I don't enjoy doing that job at all. I've also been thinking about what we learned at the zoo a few weeks ago. Zoos often help with research and conservation — but someone has to clean up after all of those animals!

That is true. You know, this reminds me of Colossians 3:23:

*Whatever you do, work at it with all your heart, as working for the Lord, not for human masters.*

Sometimes, being a good steward means that we get to do fun and exciting things as we care for God's creation. But most of the time, it means that we are faithful in doing the work that isn't all that fun.

Like taking care of our things, cleaning up after our animals, or helping to clean up garbage so that it doesn't harm any living things.

But we can do even these jobs with a good attitude as if we are doing them for the Lord.

I'm going to have a good attitude this week when I clean the cat's litter box — it's one way I'm a good steward! What about you, friend? Is there a job you can do with a good attitude, as if you were doing it for the Lord?

Look up Colossians 3:23 in your Bible. If you'd like, you can highlight this verse. Memorize Colossians 3:23 with your teacher or with a sibling.

Yippee! I can't wait to add a new page to my Science Notebook today!

Well then, what are we waiting for? I thought learning a little bit about scuba diving gear was really interesting. Can we draw a picture of a scuba diver this week?

I like that idea — and I think I saw a picture that we can use to give us an idea for our own drawings. Ah, here it is.

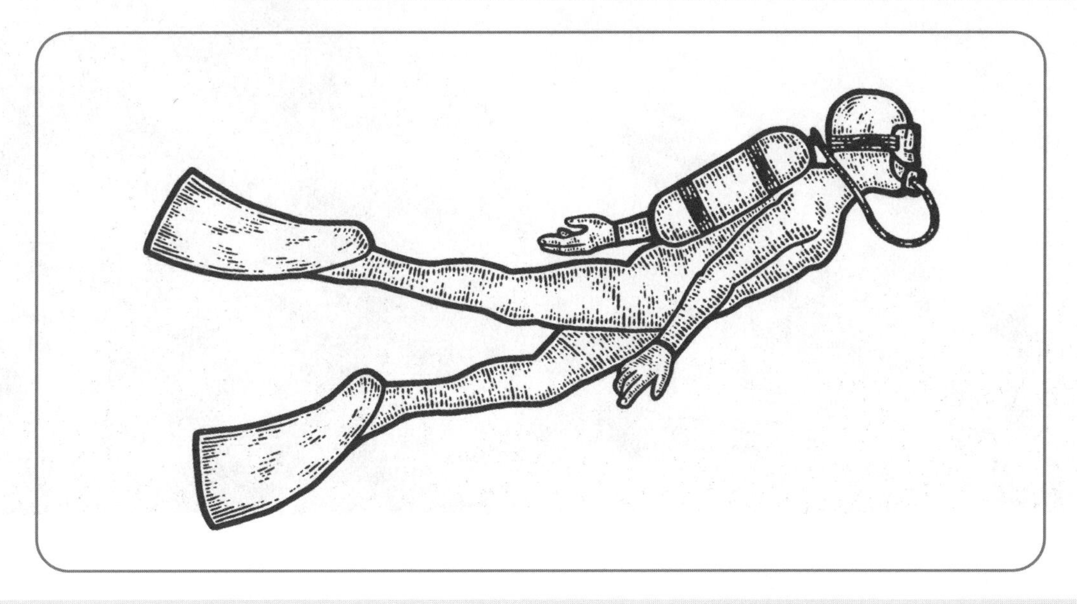

Let's get started! Remember, your drawing doesn't need to be perfect. Have fun creating your Notebook page — we can't wait to see it!

Here is how our drawings turned out. I added some little fish to mine.

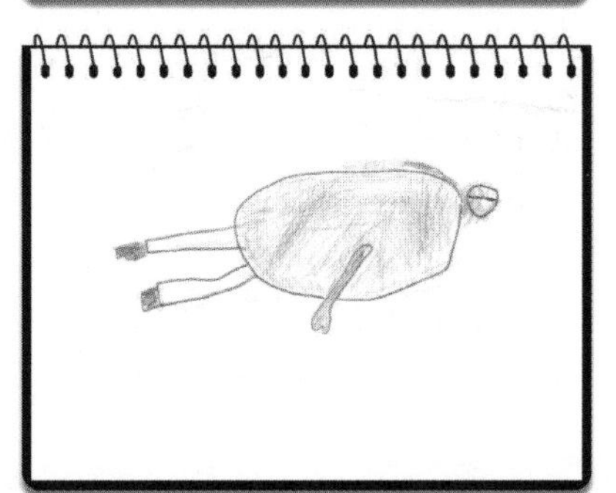

In your Notebook, write:

Scuba diving gear helps marine biologists explore the ocean.

Then draw a picture of a scuba diver.

Learning about conservation this week also reminded us that we can work as though we are working for the Lord — even when we need to do a job that isn't fun or exciting. Copy Colossians 3:23 on the back of your Notebook page as a reminder.

*Whatever you do, work at it with all your heart, as working for the Lord, not for human masters* (Colossians 3:23).

week 32

# Sharks

# Day 1

Hey there! Are you ready to start our next science adventure together? I hope so because we have more living things in the ocean to learn about!

What are we waiting for, then? This week we're going to explore my favorite fish in the ocean: sharks!

Wait, aren't sharks mammals like whales?

Good question, Ben. Remember, in order to be classified as a mammal, a living thing must have hair on its body, feed its young milk, breathe oxygen, and be a vertebrate. While sharks are vertebrates, they do not have hair on their bodies and do not feed their babies milk. This means that sharks are not mammals. Instead, they are classified as fish.

Ah, okay! I have a question, though — if baby sharks don't consume milk, what do they eat?

Baby sharks are called pups. Once they are born, the pups are ready to eat the same types of food the adult shark would eat, like fish or other small marine creatures.

Let's get back to classification now. In order to be classified as a fish, a living thing must be a vertebrate that lives in the water. A fish must also have gills to absorb oxygen from the water and fins to swim through its marine environment. Finally, fish are cold-blooded.

Ben and I learned about cold- and warm-blooded creatures in *Adventures on Planet Earth*. Let's review what we learned. God designed a mammal's body to be able to keep its temperature the same. Our bodies, for example, stay around 98.6 degrees Fahrenheit whether it is hot or cold around us. This is referred to as being warm-blooded, or **endothermic** (said this way: ĕn-dōh-thur-mĭc).

- [x] Plastic tablecloth
- [ ] Modeling clay
- [ ] Toothpick
- [ ] White acrylic paint
- [ ] Paintbrush

**Weekly materials list**

Name: ____________________

I have a way to help us remember that word. Thermic is a word that means heat, and the prefix "endo" means inside. So, endothermic means that heat is generated inside!

Thanks, Ben! On the other hand, creatures like reptiles and fish rely on the temperature of the air or water around them to regulate their body temperature. For fish, this means that if the water is cold, their body temperature will also be cold. If it is warm, then their body temperature will also be warm. This is referred to as being cold-blooded, or **ectothermic** (said this way: ěk-tōh-thur-mĭc).

The prefix "ecto" means outside — so the word ectothermic means outside heat. This can help us remember that ectothermic creatures are cold-blooded!

We're going to learn a lot about sharks in our next few adventures together. For today, though, let's review what we've learned.

1. What does the word ectothermic mean?

______________________________

______________________________

______________________________

2. What does the word endothermic mean?

______________________________

______________________________

______________________________

3. Circle the features that classify a living thing as a fish:

Invertebrate    Hair    Warm-blooded

Lives on land    Lives in water

Fins    Cold-blooded    Vertebrate

Gills    Drinks milk

We're back and ready to learn more about sharks together! I have a question to get us started today: why would we want to learn about sharks, anyway? Aren't they just mean and scary creatures, Hannah?

Well, people do like to make sharks seem like scary monsters sometimes. While some sharks can definitely be dangerous, just like tigers or bears, they are also fascinating and beautiful creatures. You may even be surprised to learn that the largest shark in the ocean is actually a gentle giant! We'll learn more about that shark soon.

First, let's learn about the anatomy of a shark. You may notice that their anatomy is similar to whales — but a big difference is that sharks have gills instead of a blowhole to breathe.

**materials needed**

- Plastic tablecloth
- Modeling clay
- Toothpick

That makes sense since sharks are fish and not mammals.

Yes, most sharks have 5–7 gill slits located on either side of their head. Like whales, sharks also have pectoral and dorsal fins. These are used to help stabilize the shark as it swims through the water.

You may also notice that a shark's tail is similar to a whale's tail in its triangular shape. However, a whale's tail is designed to move up and down as the whale swims, and a shark's tail is designed to move from side to side. A shark's tail can also be called the caudal fin (said this way: kåwd-l fĭn). Unlike a whale's tale, a shark's caudal fin is usually uneven or asymmetrical (said this way: ā-sĭ-mĕt-rĭ-kŭl). This means that one half of the tail is longer than the other half.

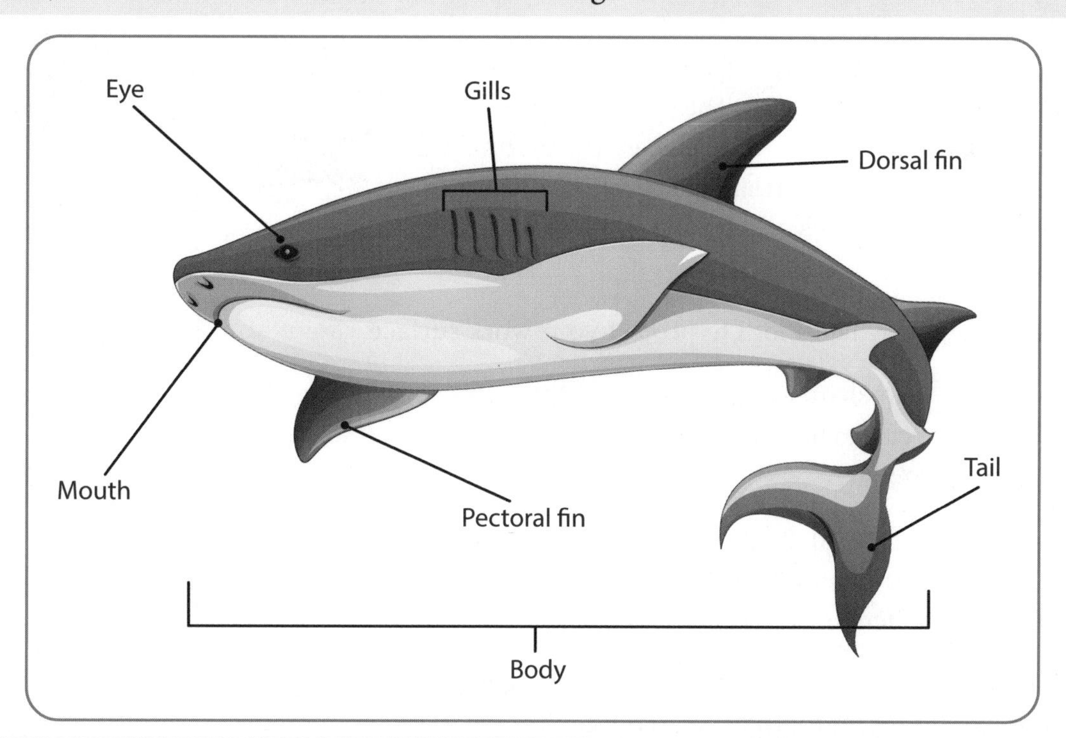


And don't forget about one of the shark's most well-known features: their triangular teeth. If you've ever seen a picture of a shark's open mouth, you may have noticed that it had rows of teeth. This is a special design that God gave to sharks — let's learn about it!

God gave you and me two sets of teeth. Our first set of teeth comes in when we are babies. Then when we are a little older, these baby teeth begin to fall out and are replaced with our adult teeth. We'll usually have our adult teeth for the rest of our lives. A shark, on the other hand, can grow several sets of teeth throughout its lifetime. This is because a shark's teeth actually aren't very strong, and they can fall out easily when the shark bites its prey.

Without the unique design God gave to sharks, it wouldn't take long before a shark had no teeth left in its mouth! However, since sharks can grow several sets of teeth in rows, if one tooth in the front row falls out, another tooth can simply move forward to take its place.

I think that is really cool! Hey, Hannah, remember when we found shark teeth on the beach during our vacation?

I sure do. It was so neat to be able to hold and feel a real tooth from a shark. I've lost the tooth that I found at the beach, but we can create a model of a shark tooth — let's go get started.

## Activity directions:

Shark teeth can be found in a variety of shapes and sizes, depending on the species of shark. The bottom, or base, of the tooth is called the root — this is where the tooth anchors to the shark's jaw. The triangular portion above the root is called the crown. Look at the edge of the shark tooth in the image:

1. Spread out a plastic tablecloth to protect your work surface.
2. Begin to form your shark tooth using about a handful of modeling clay. First, divide the clay in half. Then roll and flatten the first half to form the root of the tooth.
3. Once the root is formed, begin to form the triangular crown of the tooth and carefully smoosh the bottom of the crown into the top of the root to join them together.
4. Use the toothpick to create serrations along the edge of the crown by pressing the edge of the toothpick into the edge of the shark tooth model.
5. Allow your model to dry.
6. Once it has dried, be sure to show someone your model shark tooth!

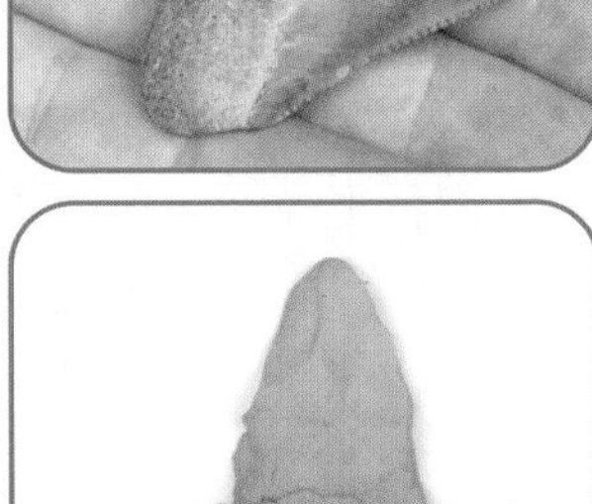

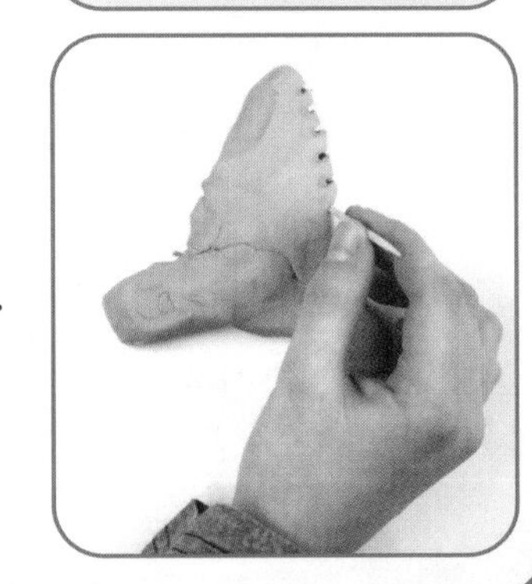

I sure had fun creating my model shark tooth, and I can't wait to show it to Uncle Gus this weekend. Hey, Hannah, you mentioned the other day that the largest shark in the ocean is actually a gentle giant. Can you tell us more about that shark?

I thought you would never ask! The largest of all sharks is called the whale shark. Whale sharks usually grow to be between 18 and 32 feet long, but they can reach up to 40 feet. Their incredible length makes them both the largest shark and the largest fish in all the ocean!

Wow! But even the largest fish in the ocean would look small compared to the 80-foot-long blue whale we learned about.

Very true. The whale shark can weigh over 40,000 pounds, and its huge mouth can span about four feet wide. You can recognize this shark by its large size and beautiful pattern of white spots and lines that help to camouflage it in the ocean water. These patterns are unique to each individual whale shark — just like your fingerprints are unique to you. I think it is amazing that God gave each one of these creatures a unique pattern!

I do too. But wait a minute, go back to the size of its mouth. A whale shark's mouth is about as wide as I am tall!

Don't worry, Ben, despite its large size, the whale shark is known as the ocean's gentle giant. In fact, divers are often able to swim right beside whale sharks because they are not a threat to humans.

Phew! So, what does this gentle giant eat?

Whale sharks are filter feeders. They eat mostly plankton and shrimp, though they may also eat small fish at times.

Name: ____________________

Hmm, the blue whale is also a filter feeder. It uses baleen plates to filter plankton from the water. But I don't see baleen plates in the whale shark's mouth. How is it able to filter its food from the water?

A whale shark opens its mouth and sucks water into it, acting kind of like a vacuum cleaner. As the water passes through its mouth and gills, plankton is caught by special gill rakers. Gill rakers function like a filter, trapping food for the whale shark to eat.

Whale sharks aren't the only type of filter-feeding shark in the ocean, though! The basking and megamouth sharks are also filter feeders. The megamouth shark was discovered more recently in 1976. Though we now know about this species, it is still very rare to see one in the ocean.

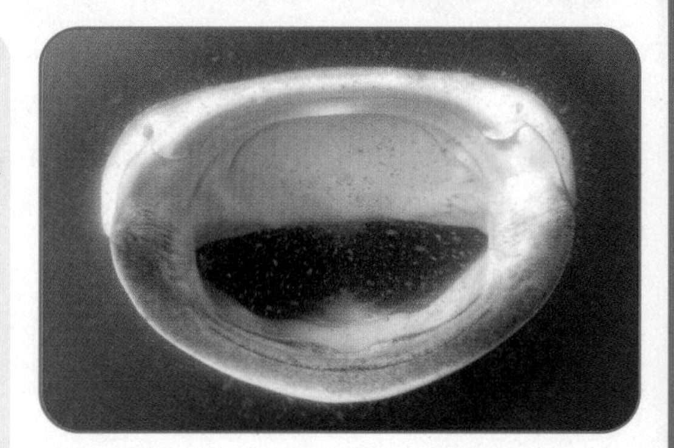

1. Imagine you are the first to discover a whale shark. How would you describe this shark to your friends?

2. A megamouth shark wasn't discovered until 1976. Do you think there are other living creatures in the ocean that we haven't discovered yet?

3. Why or why not?

Wow, Hannah. I'm beginning to see why sharks are your favorite fish in the ocean. I'm glad we're learning about them together. I can't wait to learn more next week.

I'm excited to learn more too! But first, we forgot to talk about a shark's skin the other day when we were learning about a shark's anatomy.

Sharks always look so smooth in the pictures I've seen. I really wish I could pet a shark!

Hmm, if you could pet a shark, you would need to be very careful.

Well, I know that — I'd have to watch out for their teeth.

Their teeth and their skin. A shark's skin may look smooth, but it is made of small scales called dermal denticles (said this way: dur-mŭhl dĕn-tĭ-kŭhls). Dermal denticles are like very small teeth layered to face the same direction.

A shark's dermal denticles are layered with their tips facing the shark's tail. This gives the shark a smooth body and allows water to easily pass by the shark as it swims. If you were to pet the shark from head to tail, it would feel smooth. However, if you were to pet the shark from the tail to the head, it would feel rough like sandpaper because your hand would push against the tips of the dermal denticles.

So there is more to a shark's skin than what we can see!

Yes, there is. Hmm, that reminds me of what God told the prophet Samuel in the Bible. In the Book of 1 Samuel, God had sent Samuel to the house of Jesse in Bethlehem. Samuel's job was to anoint a new king in place of King Saul. God told Samuel that one of Jesse's sons would be the next king of Israel.

At first, Samuel was looking for someone who looked like he could be a king — someone handsome, strong, and tall. Samuel thought Jesse's son Eliab was the perfect man for the job. But in 1 Samuel 16:7, God told Samuel,

*Do not consider his appearance or his height, for I have rejected him. The LORD does not look at the things people look at. People look at the outward appearance, but the LORD looks at the heart.*

When the Bible talks about our heart, it is talking about our thoughts, will, and emotions. We can make ourselves look good on the outside — we can even make it look like we are following God on the outside. But God sees deeper than that. He sees our heart on the inside, and He knows whether or not we are following Him with our whole heart. Following God with all of our heart is what matters most!

Look up 1 Samuel 16:7 in your Bible. If you'd like, you can highlight this verse. Memorize 1 Samuel 16:7 with your teacher or with a sibling.

**materials needed**

- White acrylic paint
- Paintbrush

Oh good, you're here. Do you know what time it is? It's time to add a brand-new drawing to our Science Notebook!

Yay! Let's draw a picture of a whale shark this week. I have an image right here that we can use for an example.

I can't wait to get started! But how are we going to add the white spots and stripes?

I have an idea for that — let's draw our whale shark first and color it in. Then we can use white paint to paint the spots and stripes on top of our drawing. Are you ready to get started?

Here is how our whale sharks turned out. We used colored pencils to draw and color them. Then we painted on the spots and stripes. I love how each of our whale sharks are unique — just like the real ones!

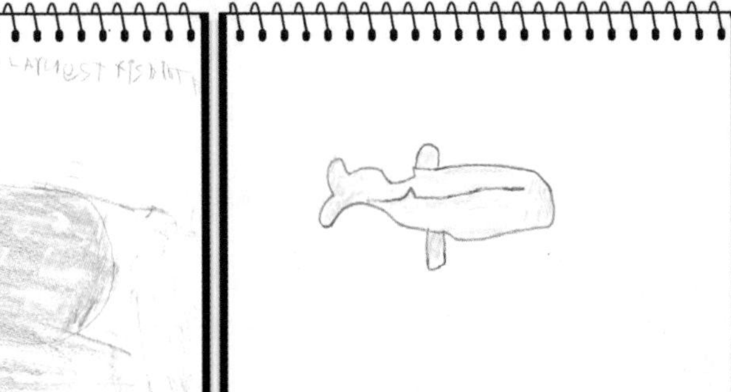

In your Notebook, write:

The whale shark is the largest fish in the ocean.

Then draw a picture of a whale shark.

Learning about sharks this week also reminded us that God sees our hearts. Copy 1 Samuel 16:7 on the back of your Notebook page as a reminder.

*Do not consider his appearance or his height, for I have rejected him. The LORD does not look at the things people look at. People look at the outward appearance, but the LORD looks at the heart* (1 Samuel 16:7).

week 33

# Sharks

# Day 2

Oh, hey there, friend, I'm glad you're back for another science adventure!

We started learning about sharks last week, and this week we're going to study three different sharks: the great white, hammerhead, and megalodon (said this way: mĕh-gŭh-lŭh-dŏn).

Ooh, I'm so excited! Let's start with one of the most well-known sharks: the great white. Pictures of great white sharks are common in books and on television. You can recognize these sharks by their gray upper body and white belly.

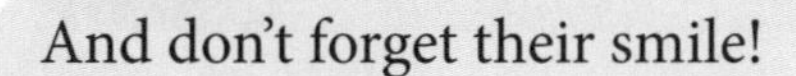

And don't forget their smile!

Oh, Ben, you're so funny.

Okay, a great white shark doesn't smile. But if one opened its mouth wide, you would see around 300 sharp teeth in rows along their large jaws.

Whoa! The great white shark can grow to be between 15 and 20 feet long. Their great size makes them the third-largest fish in the ocean — only the whale shark and basking shark are larger.

The whale shark and basking shark are filter feeders. However, the great white shark is a predatory fish. Remember, a predator is a consumer that hunts and eats other fish or animals. Great whites hunt other fish as well as marine mammals such as seals and dolphins. Great white sharks are the largest of all predatory fish in the ocean.

The great white shark has several features that make it a very skilled predator. God created sharks with the ability to detect electrical currents in the water through special sensors called the ampullae of Lorenzini (said this way: ăm-pŭh-lāy of lōr-ăn-zē-nē). We can also refer to these as electroreceptors (said this way: ĭh-lĕk-trōh-rĭ-sĕp-ters).

- [x] Onion
- [ ] Knife (adult supervision)
- [ ] Cutting board
- [ ] Tape measure

**Weekly materials list**

Wait — I know that electrical currents flow through the wires in our house. That is how we receive electricity. But why would being able to detect electrical currents help a shark?

Good question — electrical currents are actually all around you! Electrical currents are released by the cells of all living things when their heart beats or by their muscles as they move. The great white shark uses its electroreceptors to sense these electrical currents and locate its prey. In addition to electroreception, great white sharks have a strong sense of smell.

Whoa, they can smell in the water?

Yup! In fact, the great white can smell a small drop of blood in the water from over 1,000 feet away.

That's incredible — and it gives me an idea. Let's test our own sense of smell today!

## Activity directions:

ASK PARENT FOR HELP

1. Write your name and the date on your lab report on the next page. Next, write the question: How far away can I smell an onion? Write your answer in the Hypothesis section of your lab report.
2. Ask your teacher to cut an onion in half or dice it into pieces. Then go to the other side of the room or the other side of your house.
3. Carefully smell the air. Can you smell the onion at all? If not, move a little closer.
4. Move closer and closer until you detect the smell of the onion.
5. Once you can smell the onion, ask your teacher to use the tape measure to measure how close you are.
6. Write your observation in the "Things I observed" section of your report.
7. You can repeat this test — or challenge a friend or sibling! Then finish your lab report and record the results of your experiment. Is your sense of smell as strong as a great white shark?

Name ______________________ Date ______________

# Lab Report

**Question**

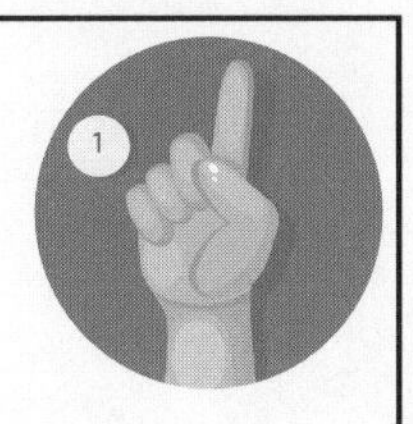


**Hypothesis**

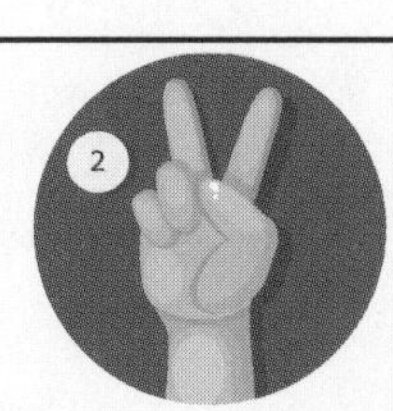


**Things I observed:**

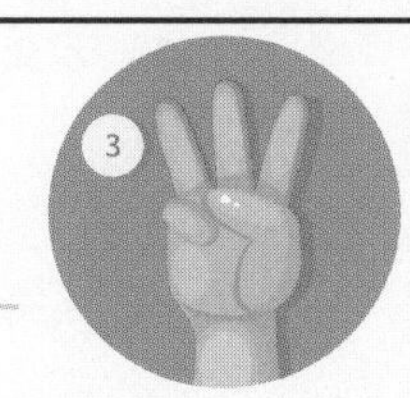


## Results

**What happened in the experiment?**

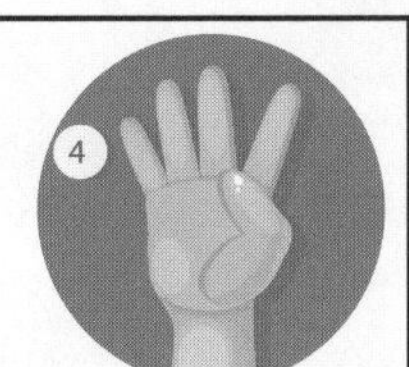


**Was my hypothesis correct?**

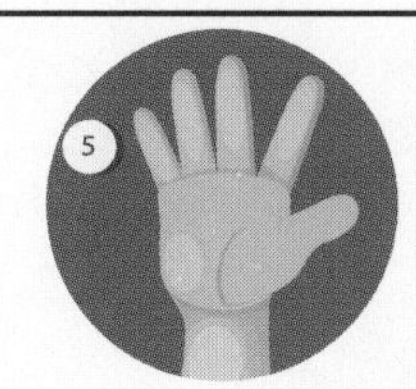

# Additional Lab Notes

Yay, I'm so glad you're here, friend! Are you ready to begin today's science adventure? We're going to learn about my favorite species of shark, hammerheads.

Ooh, hammerhead sharks are very interesting. You can recognize this species of shark by the unique shape of their heads.

That's right. A hammerhead shark's head spreads out flat and wide. The shape almost looks like a hammer — I guess that is how they got their name! The structure that creates this flat and wide shape is called a cephalofoil (said this way: sĕf-ŭh-lō-foil).

There are nine different species of hammerhead shark: the great hammerhead, scalloped hammerhead, winghead, scoophead, smooth hammerhead, Carolina hammerhead, smalleye, bonnethead, and the scalloped bonnethead. Each species has a slightly different cephalofoil shape.

Let's learn a little about the great hammerhead first. As you might have guessed from the name, this species is the largest of all the hammerhead sharks. It can grow to be around 13 feet long. The great hammerhead has a large rectangular cephalofoil and a tall dorsal fin.

Marine biologists who study hammerhead sharks believe that the unique design of their head gives them a few advantages. First, the cephalofoil gives the shark's electroreceptors space to spread out wide along the bottom of the head. This additional space for electroreceptors may help the hammerhead hunt more efficiently.

The second advantage is in their vision. A hammerhead's eyes are positioned on either side of the cephalofoil. This gives them a wide range of vision — some species can even see all the way around behind them!

But that's not all. The great hammerhead shark uses its keen vision, sense of smell, and electroreceptors to hunt stingrays. Stingrays often bury themselves in sand in order to hide — but they can't hide for long! Once the hammerhead has found a stingray, it can use its large flat head to pin the ray down so that it can't escape.

Another species of hammerhead is the scalloped hammerhead. Most scalloped hammerheads grow to be between 6–8 feet long. Scalloped hammerheads are quite similar to great hammerheads, except that their cephalofoil has a wavy or scalloped edge.

Name: ______________________

Not all species of hammerhead sharks are large. The bonnethead (said this way: bŏn-nĕt hĕd) shark only grows to 30–48 inches long. A bonnethead's cephalofoil is shaped like a rounded baby bonnet. These small hammerheads hunt small fish, crabs, and shrimp.

I love seeing God's creativity on display within the hammerhead species!

(CC BY 3.0 AU)

1. This is a winghead shark. If you were a marine biologist, how would you describe this species of hammerhead to someone else?

2. This is a scoophead shark. How would you describe this species of hammerhead to someone else?

You know, Hannah, I wasn't sure I would enjoy learning about sharks at first. But I've had a lot of fun! Let's review what we've learned so far. First, we talked about what features classify a shark as a fish and about shark anatomy. After that, we were able to learn a little about the whale shark, great white, and hammerheads.

And we've only scratched the surface! There are over 500 different species of sharks in the ocean.

Whoa, that's a lot of different species — and that's just sharks! There are thousands of other species of life in the ocean. It reminds me of what God said in Genesis 1:20,

*And God said, "Let the water teem with living creatures, and let birds fly above the earth across the vault of the sky."*

Teem is a word that means to be full of. The ocean certainly does teem with living creatures!

During our adventures in marine biology, we've learned a lot about the creatures that we can observe alive today — this is part of observational science. Remember, observational science is science that we can see, experience, or observe. This type of science is observable, testable, and repeatable.

Marine biology doesn't only deal with observational science, though. Did you know that sometimes sailors or marine biologists discover evidence of creatures that once lived in the ocean?

You mean creatures that are extinct now?

Yes. Just like we find evidence of dinosaurs and woolly mammoths on land, we also find evidence of extremely large sea creatures that once lived in the ocean. One such creature is called megalodon (said this way: mĕh-gŭh-lŭh-dŏn).

Oh, I've read about this one! We're able to learn a little bit about dinosaurs through their fossilized bones. However, sharks are a little different. Their skeletons are made of cartilage instead of bones.

Artist rendition of a megalodon chasing a kentriodon.

Feel the top part of your ear — it's also made of cartilage!

Cartilage isn't hard like bone, and it doesn't fossilize well. This makes it very, very rare to find parts of a shark that have been fossilized. But we do find fossils of their teeth. The teeth we've found from a megalodon can be as big as a man's hand! Scientists estimate that this shark would have been about 50 feet long. It must have been an incredible sight.

Name: ______________________

However, no one has ever found a live megalodon that we can study today. We cannot observe its color, behavior, diet, speed, or even how long it lived. When we learn about creatures like megalodon, we're exploring historical science. Historical science cannot be observed, tested, or repeated.

Instead, we must develop theories about the creature based on the evidence we find and facts we know about similar creatures. Remember, a scientific theory is a logical way to explain what we see or to answer a question based on evidence and facts.

As we learn about creatures that lived in the past, it's important to remember two things. First, we need to remember that we're exploring historical science that cannot be observed, tested, or repeated. Second, no matter what we are exploring in science, it is important to always compare the theories scientists have to what the Bible teaches us.

The teeth we find from megalodon are similar to the teeth we find from great white sharks today, but much, much larger. Scientists believe that the megalodon may have looked like a giant great white shark. What do you think this incredibly huge shark looked like? Draw a picture of what you imagine it was like below.

Comparing the size difference between a megalodon and a great white tooth.

Hello, friend! Did you have fun drawing what you imagine megalodon looked like? Ben and I definitely did!

I've been thinking about what Ben mentioned yesterday about Genesis 1:20. The ocean is certainly full of all kinds of amazing creatures. Like we talked about yesterday, we find evidence of other incredible marine creatures like kronosaurus and the plesiosaurus that lived in the past. Did you know that the Bible also talks about one incredible sea creature named Leviathan? Psalm 104:24–26 says,

*How many are your works, Lord! In wisdom you made them all; the earth is full of your creatures. There is the sea, vast and spacious, teeming with creatures beyond number — living things both large and small. There the ships go to and fro, and Leviathan, which you formed to frolic there.*

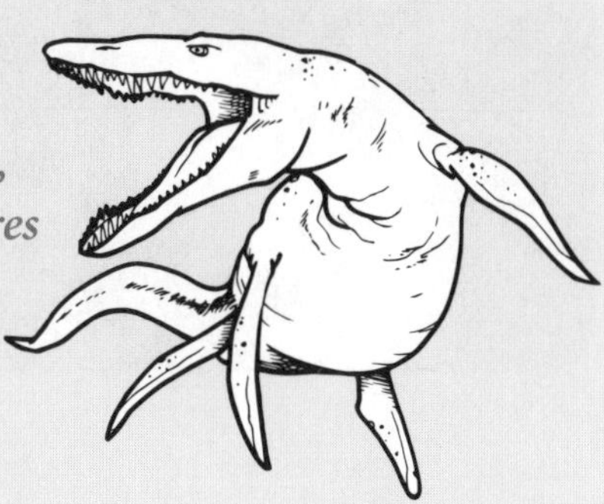

Hmm, that doesn't tell us much about this sea creature.

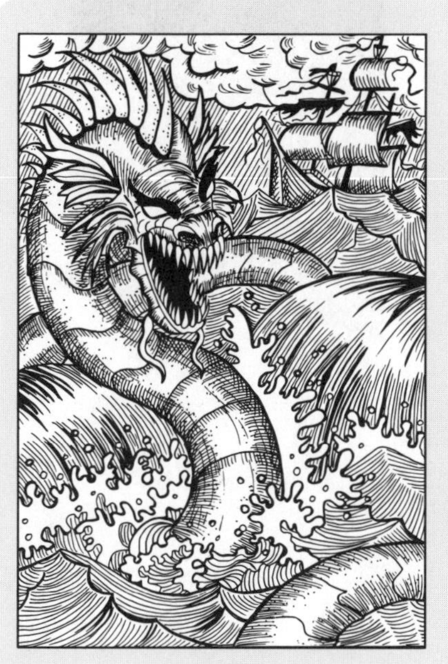

No, but we learn more about it in the Book of Job. In chapter 41, God Himself tells Job about this fearsome creature:

*Who dares open the doors of its mouth, ringed about with fearsome teeth? Its back has rows of shields tightly sealed together; each is so close to the next that no air can pass between. They are joined fast to one another; they cling together and cannot be parted* (Job 41:14–17).

These verses describe a creature with fearsome teeth and strong scales. Later in chapter 41, God says that sword and spears do not scare this creature — He says that these things have no effect on Leviathan! But verses 18–21 get even more interesting:

*Its snorting throws out flashes of light; its eyes are like the rays of dawn. Flames stream from its mouth; sparks of fire shoot out. Smoke pours from its nostrils as from a boiling pot over burning reeds. Its breath sets coals ablaze, and flames dart from its mouth.*

Whoa, it sounds like God is describing a fire-breathing sea creature.

It sure does! Now, some believe that maybe God is being poetic about this amazing sea creature. However, we do know that certain living things are able to use chemicals to make light or spray incredibly hot liquid when they are threatened.

You're talking about lightning bugs and the bombardier beetle, right?

Yup! In verse 33, God says this about Leviathan,

*Nothing on earth is its equal — a creature without fear.*

Though we don't have any fire-breathing marine creatures alive today, it is absolutely possible that God created one that lived in the past. Nothing on earth would have been its equal for sure! This creature certainly would have proclaimed God's majesty and awesome power. Be sure to read Job chapter 41 together with your family. What kind of creature do you think Leviathan was based on what the Bible says?

You can explore other amazing sea creatures in *Dragons of the Deep*, available through Master Books (optional). Read Job chapter 41 together as a family and talk about what you think it would have been like to be able to see this amazing creature.

Did you bring your art supplies with you today, friend?

I hope so because it's time to add a new page to our Science Notebook!

We learned about great white sharks, hammerheads, and megalodon this week. I was thinking that it would be fun to draw a picture of a hammerhead shark! Here is an image that we can use for an example.

Or if you'd like, you can use one of the pictures of a hammerhead shark from our lesson earlier this week to give you an idea for your drawing.

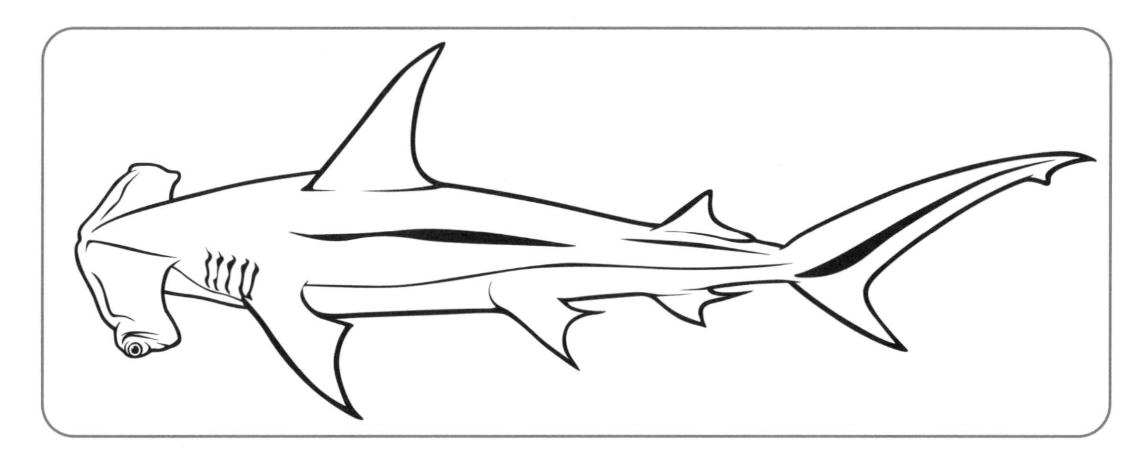

Here is how our hammerhead shark drawings turned out. Have fun drawing your hammerhead shark, friend — and don't forget to show it to someone else!

In your Notebook, write:

There are nine different species of hammerhead shark.

Then draw a picture of a hammerhead shark.

Learning about sharks this week also reminded us that God created many incredible creatures, like whales, sharks, and even Leviathan, to live in the ocean. Copy Psalm 104:26 on the back of your Notebook page as a reminder.

*There the ships go to and fro, and Leviathan, which you formed to frolic there* (Psalm 104:26).

week 34

# Arctic Ocean

## Day 1

Can you believe it's time for our next science adventure? We've been exploring marine biology, and it has been so much fun!

For sure! We've already learned about waves, tides, tide pools, algae and seagrass, the marine food chain, coral reefs, whales, conservation, and sharks.

Whew, we've sure explored a lot of different things. It makes me wonder, is there anything else for us to learn about in marine biology?

Absolutely! There is still so much more we could study in this field — more than we can possibly fit into our short adventures together. Actually, that is one of the things that amazes me most about God's creation. No matter how much we learn about it, there is still so much more we can learn.

Now it's time to begin our next science adventure together, and you might want to grab your winter jacket this time.

Wait — now I'm confused. Hannah, you just said that there is still so much more that we can explore in marine biology. Why would I need a winter jacket?

Because we're heading to the arctic now, the Arctic Ocean! You may remember from our previous adventures that over 70% of the earth is covered in water. Though all of the oceans are connected, we divide them into five oceans.

I remember — there is the Atlantic Ocean, Indian Ocean, Pacific Ocean, Southern Ocean, and the Arctic Ocean.

**Weekly materials list**

- [x] Water and ice cubes
- [ ] Drinking glass
- [ ] Shoebox
- [ ] Plastic tablecloth
- [ ] Light and dark blue acrylic paint
- [ ] Paintbrush

Name:

Correct. The Arctic Ocean is the smallest of all the oceans. It is located high in the northern hemisphere of the earth. When you hear the word "arctic," you probably think of ice and snow — and you'd be right! The arctic is home to glaciers, icebergs, and sea ice.

I would imagine that would make this ocean very difficult to explore.

For sure! The thick ice and frigid temperatures make studying living things in the arctic quite difficult. There is a lot that marine biologists simply don't know or understand yet about life in the Arctic Ocean.

Well, I'm excited to explore what we do know about the arctic with you in our next few adventures together!

1. What is something you would like to learn about the Arctic Ocean?

2. Why do you think it is difficult for marine biologists to study life in the cold Arctic Ocean?

Hello, friend! Before we begin exploring the creatures that live in the arctic, we're going to learn more about the arctic environment. Ready to get started?

**materials needed**

- Ice cubes
- Drinking glass
- Water

The Arctic Ocean is the world's coldest ocean, and it is a very harsh environment. The average temperature of this ocean is around 28° Fahrenheit. But if you think that is cold, you should feel the frigid winter air, which can average around -30° Fahrenheit — brrr!

I feel cold just thinking about it!

These cold temperatures contribute to the formation of glaciers, icebergs, and sea ice. A **glacier** (said this way: glāy-shur) is a very large area of ice. In fact, a glacier can extend over 200 miles long. In an environment like the arctic, the snow that falls each year does not fully melt. Glaciers are formed over time by layers of snow and ice that have not melted. Each new layer presses down and compacts the layers beneath it to form the glacier.

Glaciers slowly creep along the land due to their icy base and gravity. When a glacier reaches the ocean, chunks of it may break away. The massive chunks of ice float in the water, and we call them icebergs. Only a portion of the iceberg is visible at the surface of the water. Most of the iceberg is found beneath the surface of the waves.

Finally, sea ice is formed on the surface of the ocean as water freezes in the cold temperatures. Sea ice is very important to some of the living things we'll be learning about soon.

Hmm, I have a question. We know that ice is made of water that has frozen. So, why does ice float on water?

Good question, Ben. The answer lies in density. **Density** (said this way: dĕn-sĭ-tē) is the measurement of how much matter is in an object and how much space that matter takes up. Objects with a high density, or matter that is pressed into a small area, will sink in water. For example, a rock has matter that is compacted, or pressed into a small area. This causes the rock to sink in water because it is more dense than the water.

On the other hand, an object with a lower density than water will float. As water freezes, the molecules arrange themselves in such a way that the ice becomes slightly less dense than the water. This means that the water around it is denser than the ice, which causes the ice to float at the surface.

This feature of water reveals God's incredible wisdom because it allows ice to form at the top of ponds, lakes, and the ocean. Since ice forms at the top of water, marine life is able to swim underneath it rather than be frozen in the ice. Let's observe some ice together!

## Activity directions:

We're going to put ice cubes into the bottom of a cup and then fill the cup with water. What do you think will happen to the ice cubes — will they remain at the bottom of the cup or float to the surface?

ASK PARENT FOR HELP

1. Write your name and the date on your lab report on the next page. Next, write the question: Will the ice cubes stay at the bottom of the cup or float to the surface?
2. Write your answer in the Hypothesis section of your lab report.
3. Place 2–3 ice cubes in the bottom of the glass. Then fill the glass with water.
4. Write your observation in the "Things I observed" section of your report.
5. Finish your lab report and record the results of your experiment.

**Bonus!** Fill a plastic container with water and freeze it to create a larger ice cube. Place the larger ice cube into a sink or bathtub full of water. What do you observe?

Name ______________________________ Date ____________________

# Lab Report

## Question

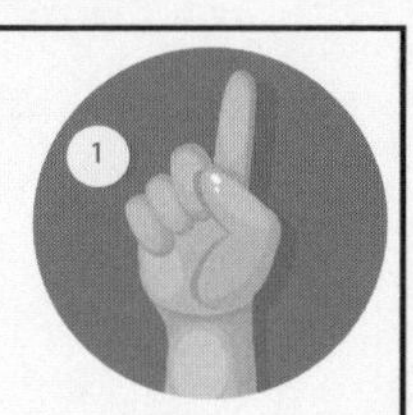


## Hypothesis

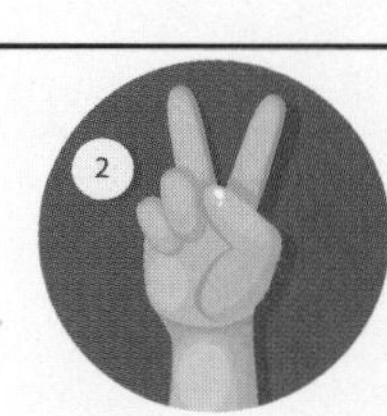


## Things I observed:

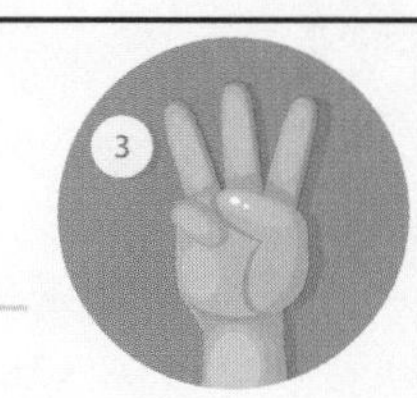


## Results

### What happened in the experiment?

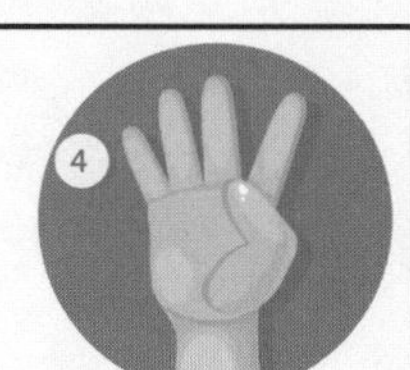


### Was my hypothesis correct?

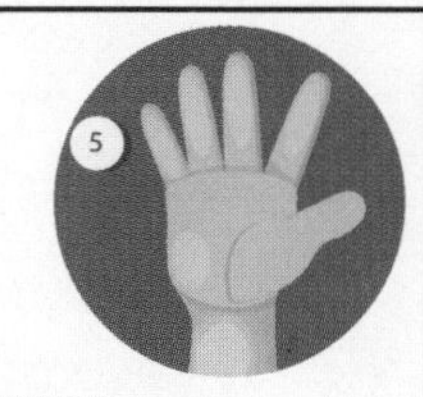

## Additional Lab Notes

Welcome back, friend. Are you ready to dive deeper into our exploration of the Arctic Ocean? Let's start with one of my favorite arctic creatures: the polar bear!

Wait a minute, Hannah — I thought we were exploring marine biology. Polar bears live on land!

**materials needed**

- Shoebox
- Plastic tablecloth
- Light and dark blue acrylic paint
- Paintbrush

Polar bears are actually classified as marine mammals. This is because they spend much of their time on the sea ice and rely heavily on the Arctic Ocean. In fact, the polar bear's Latin name in taxonomy is *Ursus maritimus* (said this way: rr-sŭs mair-ĭ-tī-mŭs), which means "sea bear."

Oh, okay, that makes sense!

One thing that amazes me about God's creation is that He gave living creatures the ability to adapt. **Adapt** (said this way: ŭh-dăpt) means to adjust or change for certain conditions or a particular environment. The polar bear has adapted for life in the harsh arctic environment in a couple of ways.

The first way is their thick fur coat. Maintaining body temperature is a very important feature for creatures living in the frigid arctic. The polar bear's fur coat has two layers of fur to keep it warm. The undercoat is the first layer, and it is made of thick, short hair that insulates the bear. The outer layer of fur is made of longer hair. The outer layer acts as a guard or shield over the undercoat to protect it from the harsh weather and cold water.

Now I know why polar bears look so fluffy and white — it's all that fur!

Actually, a polar bear's fur is transparent — it doesn't have a color.

Wait — their fur is white, isn't it?

Ah, that is one of the most interesting adaptations of the polar bear! Their fur is transparent, or clear. Their transparent fur allows sunlight to pass through to their black skin. The color black absorbs the most heat from sunlight. This design helps keep the polar bear warm. However, we see their fur as white because of the way light is reflected.

Wow, God perfectly designed this bear for its environment!

He sure did. The polar bear also has a very thick layer of fat under its skin. This fat layer can be over four inches thick, and it also helps to insulate the bear from the cold. Of course, in order to maintain that layer of fat, a polar bear must also eat a lot of fat.

Polar bears are the apex predator of the arctic. They primarily hunt ringed seals, which supply them with a lot of fat. We'll learn more about ringed seals later on. Polar bears also hunt other species of seal, walruses, narwals, and beluga whales.

I can't wait to learn more about ringed seals. I have a surprise for you now, though — Mom said we could create an arctic biome! Let's get started on it together.

## Activity directions:

ASK PARENT FOR HELP

1. Spread out a plastic tablecloth to protect your work surface.
2. Paint the bottom of your shoebox dark blue. This will be our Arctic Ocean.
3. Rinse your paintbrush.
4. Paint the rest of the inside of your shoebox light blue. This will be our arctic sky.
5. Set the biome in a safe place and allow it to dry.

6. Rinse out your paintbrush — we'll continue working on our biome next week!

I've enjoyed learning about the arctic and polar bears this week. I can hardly wait to continue our arctic adventure next week.

There is so much here that we can learn about — we didn't even get to finish learning about polar bears yesterday!

Hmm, we have a little bit of time today. Tell us more about polar bears, Hannah.

The polar bear is the largest of all the bear species in the world — they can stand 6–9 feet tall and weigh over 500 pounds. That's a lot of weight on the ice! A polar bear's paw can be around 12 inches wide. Their wide paw spreads out their weight, which enables them to walk over the snow and ice without falling through.

But that's not all. The sea ice can also be quite slippery. The bottom of a polar bear's paw is rough, which allows it to grip the ice.

That way, the polar bear doesn't slip and hurt itself!

Yup! Isn't it amazing how God designed the polar bear with the ability to thrive in such a harsh environment? It reminds me of Psalm 121:1–3,

*I lift up my eyes to the mountains — where does my help come from? My help comes from the Lord, the Maker of heaven and earth. He will not let your foot slip — he who watches over you will not slumber.*

Hmm, I'm confused by that verse, Hannah. Sam spilled water on the floor the other day, and I slipped on it. I fell hard and even hurt my elbow! But the verse says that God will not let your foot slip . . . what does it mean?

I'm glad you asked, Ben. The Book of Psalms is full of poetry. Poetry in the Bible often gives us a picture of our relationship with God — but it isn't always literal. Literal means that the words mean exactly what they say.

So does that mean that the verse is giving us a picture of God's care in words?

Exactly! "He will not let your foot slip" doesn't mean that you will never actually slip and fall down. Instead, it's a picture of God's protection and care over us. He is ever-present, and He does not slumber or sleep. Whenever you see a polar bear now, you can be reminded of these verses!

Look up Psalm 121:1–3 in your Bible. If you'd like, you can highlight these verses in your Bible. Memorize Psalm 121:1–3 with your teacher or with a sibling.

We're back and ready to add a new page to our Science Notebook!

Yippee! Can we draw a picture of a polar bear this week?

You read my mind, Ben. I have a picture of a polar bear right here so we can use it to give us an example for our own drawings.

Here is how each of our drawings turned out. My polar bear is hunting a ringed seal under the ice. Isn't Sam's polar bear cute? Sam said that his polar bear is eating a plant, like it would have in God's creation before sin entered.

Have fun drawing your polar bear, friend!

In your Notebook, write:

Polar bears are well adapted for life in the arctic.

Then draw a picture of a polar bear.

Learning about polar bears this week also reminded us that God is ever-present and cares for us. Copy Psalm 121:1–3 on the back of your Notebook page as a reminder.

*I lift up my eyes to the mountains — where does my help come from? My help comes from the Lord, the Maker of heaven and earth. He will not let your foot slip — he who watches over you will not slumber* (Psalm 104:26).

week 35

# Arctic Ocean 2

Day 1

I'm really excited to start our adventure today, and I'm glad you're here, friend! Can we learn about ringed seals first, Hannah?

Absolutely. Ringed seals are the most common species of seal in the arctic. They get their name from the pattern of circles and rings on their fur. Ringed seals grow to be about five feet long and can weigh around 150 pounds.

Hmm, I'm guessing that ringed seals are also classified as marine mammals?

You're right! Ringed seals are mammals that spend much of their time in the arctic water. They can swim at six miles per hour and dive over 150 feet deep.

Wow, they are excellent swimmers. But as mammals, ringed seals need oxygen to breathe. How long can they hold their breath, Hannah?

A ringed seal can hold its breath for 45 minutes, which is pretty amazing. But the Arctic Ocean is a dangerous place, and the sea ice can freeze over while the ringed seal is deep under the surface of the waves.

And that's not good — if a marine mammal can't break up through the sea ice, then it isn't able to breathe air.

The ringed seal is well equipped for life in the arctic waters, though. This seal has thick, sharp claws at the end of its front flippers. It uses these claws to scrape through the ice and create breathing holes. Ringed seals are known to maintain several different breathing holes through the thick arctic ice.

## Weekly materials list

- [x] Plastic tablecloth
- [ ] Cotton balls
- [ ] White glue
- [ ] Utility knife (adult supervision)
- [ ] Pencil
- [ ] Clear or silver glitter glue
- [ ] ½–1 inch thick Styrofoam™ panel or block
- [ ] Small model arctic marine mammals such as polar bear, seal, beluga, narwhal, or bowhead whale

**Name:** ____________________

That sounds like a lot of work. Why does the seal maintain several different breathing holes?

Well, remember how we talked about polar bears hunting ringed seals? Polar bears need to maintain their body fat. Ringed seals have a thick layer of fat, or blubber, under their skin to keep themselves warm. This makes them ideal prey for polar bears.

Polar bears are excellent hunters with a strong sense of smell. They use their sense of smell to hunt for a ringed seal's breathing hole, and then they patiently wait for the seal to come up to breathe. Sometimes, the ringed seal will blow bubbles through its breathing hole to see if a polar bear is waiting for it. If a polar bear reaches through the breathing hole in response to the bubbles, then the seal knows to use one of its other breathing holes.

That makes sense. And if one hole closes over in deep ice, the ringed seal can hurry to another hole that may not have closed over yet.

Exactly! We're out of time for today, but I'm looking forward to talking more about marine life in the arctic tomorrow.

1. What features help the ringed seal survive in the Arctic Ocean?

2. How can ringed seals check their breathing holes for a polar bear?

Oh good, you're back! I have a question to get us started today. Food in the arctic can be scarce, and the creatures that live here must often go long periods of time — even months — without food. What do ringed seals find to eat, Hannah?

Great question. Ringed seals are carnivores. They eat mostly shrimp, krill, and arctic cod.

Wait, I read about arctic cod in one of my books. The arctic cod can also be called polar cod, and it's the most common species of fish in the arctic. Arctic cod grow to be about 11 inches long, and they eat zooplankton. Do you remember when we learned about the classification of fish, Hannah?

I think so . . . in order to be classified as a fish, a living thing must be a vertebrate that lives in the water, must have gills to absorb oxygen, fins to swim, and be cold-blooded.

Right! Remember, the average temperature of the Arctic Ocean is 28° Fahrenheit. As you can imagine, this freezing temperature makes it very difficult for cold-blooded fish.

Its body could freeze!

However, God designed the arctic cod to have a special compound in their blood called antifreeze glycoprotein (said this way: ăntī-frēēz glī-kōh-prō-tēēn). This compound keeps the fish from freezing and allows it to survive in the arctic. Arctic cod is an important source of food for ringed seals, arctic birds, whales, and the Greenland shark.

Wait — there is a shark in the arctic too?

There is! The Greenland shark is also able to survive in the arctic water temperature due to a chemical called trimethylamine oxide (said this way: trī-mĕh-thŭh-lŭh-mēēn ŏx-īd), or TMAO for short. TMAO helps to keep the shark's body from freezing and also helps it to withstand the pressure deep underwater. The high amount of TMAO found in the Greenland shark's flesh makes the shark toxic for humans and other mammals to eat unless the meat is specially prepared to remove this chemical.

Whoa, that is really interesting. What does the Greenland shark eat?

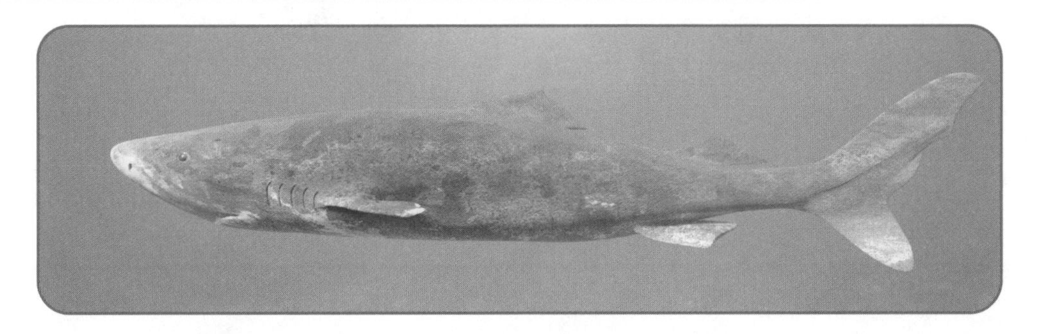

Name:

Whatever it can find! The Greenland shark is a carnivore. It preys on fish, seals, and squid. It is also a scavenger. Scavengers eat creatures that have already died.

Marine biologists still have a lot to learn about the Greenland shark. It lives deep within the frigid arctic water and hasn't been studied very much. However, scientists believe that this shark can have an incredibly long lifespan of around 400 years.

Wow, I'm excited to learn even more about arctic creatures tomorrow!

1. Think about what you've learned about ringed seals in our last two adventures. Then pretend you are a marine biologist who studies ringed seals. Describe ringed seals in 3–4 sentences.

2. How do antifreeze glycoprotein and TMAO help fish in the arctic?

Welcome back! We've been exploring the Arctic Ocean together lately. We've already learned about the conditions of the arctic, polar bears, ringed seals, arctic cod, and the Greenland shark. Today, I thought it would be fun to talk about the whales of the arctic.

Ooh, I like that idea. Can we start with the bowhead whale?

Sure! The bowhead whale is a baleen whale. It can grow to be 60 feet long and has baleen plates that can be around 13 feet long. The bowhead whale is a filter feeder that eats krill, copepods (said this way: kōh-pŭh-pŏds), and small fish.

As a marine mammal, the bowhead whale needs to be able to maintain its body heat in the cold arctic water. It has a very thick layer of blubber that insulates it from the cold. A bowhead whale can have blubber that is 18 inches thick. This enables the bowhead to spend the whole year in and around the Arctic Ocean.

One of my favorite features about this whale is that it has very strong bones in its head. A bowhead whale uses its strong skull to break through the sea ice in order to breathe. Though we still have a lot to learn about bowhead whales, scientists believe these whales can live to be around 200 years old. The arctic is home to some incredibly designed creatures!

Definitely. I think we have just enough time left today to talk about another whale that migrates to the Arctic Ocean: the beluga whale.

Ooh! Beluga whales are easily recognized by their white skin and the shape of their large forehead.

Their large forehead is called a melon. The beluga's melon helps them echolocate as well as communicate. Many species of whales live alone, but beluga whales are very social mammals. Belugas live together in groups called pods, like dolphins. They communicate with each other through many different chirps, clicks, whistles, and squeals.

Name: ______________________________

One feature beluga and bowhead whales have is that they do not have a dorsal fin. In the cold arctic water, heat would escape from their bodies through the dorsal fin. A tall dorsal fin would also be injured easily as the whale swam under the sea ice. How amazing that God designed His creation with the ability to adapt!

We've learned about several different marine mammals and fish that live in the Arctic Ocean. Use each clue to unscramble the name of the mammal or fish.

1. This marine mammal has blubber that can be 18 inches thick: **odwheba elawh**

______________________________

2. This marine mammal's Latin name means "sea bear": **parlo erab**

______________________________

3. This fish is an important source of food for marine mammals in the arctic: **olpar doc**

______________________________

4. Trimethylamine oxide keeps this fish's body from freezing: **laGrenned kashr**

______________________________

5. This whale lives in a pod and communicates through clicks, chirps, whistles, and squeals: **uablge elawh**

______________________________

6. This marine mammal has sharp claws to dig through the sea ice: **drineg asel**

______________________________

We're so close to finishing our science adventures together for this year.

And what an adventure it has been — I'm so glad we were able to explore God's creation together through chemistry and marine biology!

I've enjoyed learning about arctic marine mammals and fish this week. In our earlier adventures, we learned that God created all living things. He created them to live in a perfect world — one that was free of sin and death.

But in His mercy, God also designed creation with the ability to adapt to the conditions of an imperfect world. We learn in Genesis chapter three that creation changed after Adam and Eve sinned. Death, thorns, thistles, and suffering entered creation.

Genesis also tells us about a worldwide disaster — the Flood of Noah's day. This Flood would have greatly changed the conditions on the earth.

God gave His creation the ability to adjust and adapt. His design means that there are incredible creatures all over the earth to show us God's wisdom.

We learned at the beginning of our adventures together that science teaches us more about God and our relationship with Him. All that we've learned together reminds me of Psalm 36:5–7:

*Your love, LORD, reaches to the heavens, your faithfulness to the skies. Your righteousness is like the highest mountains, your justice like the great deep. You, LORD, preserve both people and animals. How priceless is your unfailing love, O God! People take refuge in the shadow of your wings.*

I'm so grateful that God preserves both people and animals — even in our imperfect world. Because of His mercy and wisdom, we are able to study His creation and learn more about Him.

I get excited each time I learn about something He has designed in creation. What about you, friend? Do you see God's wisdom and mercy on display in His creation? Be sure to talk to your family about your favorite part of creation — and what it has taught you about God.

Look up Psalm 36:5–7 in your Bible. If you'd like, you can highlight these verses in your Bible. Memorize Psalm 36:5–7 with your teacher or with a sibling.

Usually we would add a new page to our Science Notebook today, but instead we are going to finish creating our model arctic biome. Are you ready? Let's get started!

## Activity directions:

**materials needed**

- Plastic tablecloth
- Cotton balls
- White glue
- Utility knife (adult supervision)
- Pencil
- Clear or silver glitter glue
- ½–1 inch thick Styrofoam™ panel or block
- Small model arctic marine mammals such as polar bear, seal, beluga, narwhal, or bowhead whale

1. Spread out a plastic tablecloth to protect your work surface.

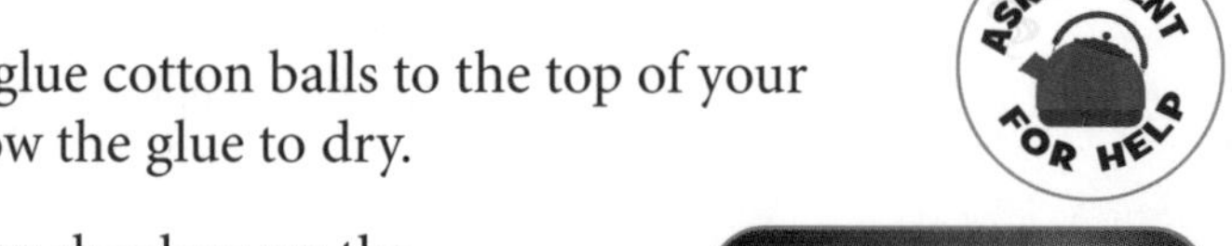


2. To create clouds, glue cotton balls to the top of your arctic biome. Allow the glue to dry.
3. Once dry, place the shoebox on the Styrofoam™ panel and use the pencil to trace the edge of the shoebox.
4. Ask your teacher to use the utility knife to cut along the inside of the line you traced.
5. Make sure the cut Styrofoam™ fits inside your box. If it doesn't, ask your teacher to trim it until it does.

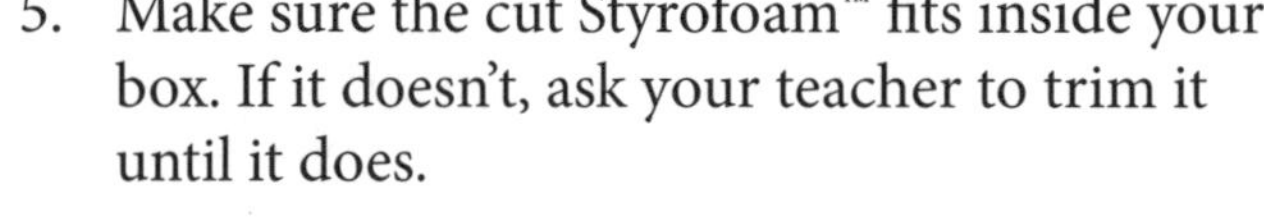

6. Now let's create our glacier! Break the cut piece of Styrofoam™ to the height you would like it to be in your shoebox. Then place it in the back of your shoebox. If it doesn't stay in place, you can glue it in.

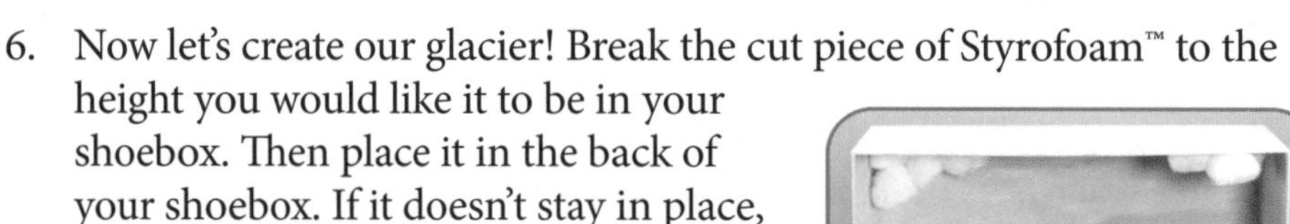

7. Break smaller pieces of Styrofoam™ and place them in the bottom of the shoebox — this is our sea ice.
8. Optional: you can use the glitter glue on the glacier or sea ice to make them sparkle like snow in the sunlight.
9. Add the small model arctic creatures to your biome.
10. Share your Arctic Ocean model with your family. Be sure to tell them about the creatures you learned about, as well as the ways God designed them to adapt.

**Bonus!** Take a picture of your Arctic Ocean model and ask your teacher to help you print it out. Then tape or glue the picture on the next page in your Science Notebook. Write **My Arctic Ocean Model** at the top of the page.

# Review

Day

Hello, fellow science adventurer! Can you believe we've already reached our last science adventure together in this course?

Time flies when you're learning about God's creation and having fun!

That's for sure. We've explored so much together this year, I thought it would be fun to look back through our Science Notebooks and remember all the different things we've learned about.

Great idea, Hannah! Grab your Science Notebook, friend, and we'll look through our drawings together. The very first drawing in our Notebooks was of the earth, one of God's great creations. In that first adventure, we learned that science is the pursuit of knowledge and understanding about God's creation through an organized process. Science helps us to ask questions, test our ideas, and share what we've learned with others — just as Psalm 9:1–2 says:

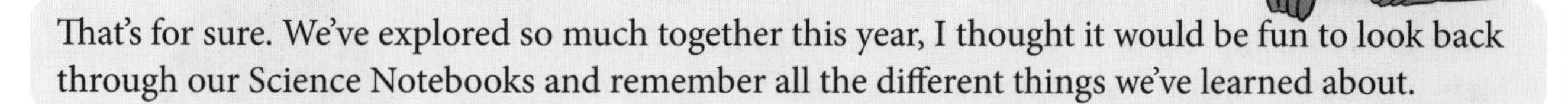

*I will give thanks to you, LORD, with all my heart; I will tell of all your wonderful deeds. I will be glad and rejoice in you; I will sing the praises of your name, O Most High.*

I love that verse. Turn to the next drawing in your Notebook. We drew a vanilla flower and bean to help us remember our second adventure!

Yuck, I definitely remember the vanilla extract. We learned that our senses and logic can't always be relied upon as we study science — and that is where the scientific method comes in! Flip over to the next drawing. We learned the steps of the scientific method in our third adventure:

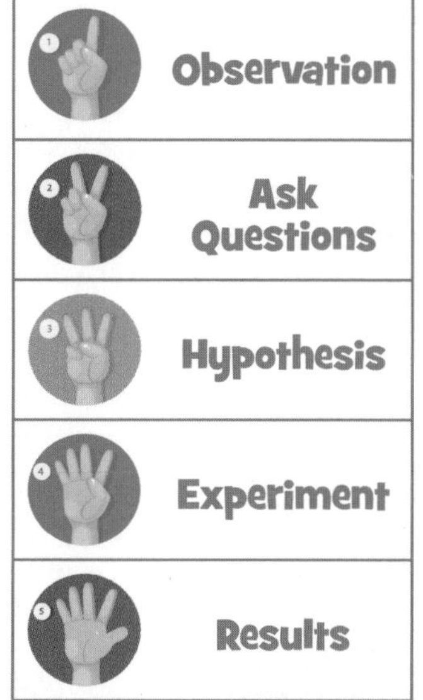


The scientific method helps us study science in an organized way as we ask questions, test our ideas, and develop conclusions. During our adventure that week, we also learned that science can help us understand more about God's creation, but it cannot give us complete truth — only God and His Word can do that. Though science may change at times, we know that God and the truth He reveals in Scripture never, ever changes.

Name: ______________________________

That is so important to remember, no matter what we are studying! All right, now turn to the drawing of a lab notebook. In our fourth adventure together, we learned about lab reports. Lab reports help us to document our experiments and observations so that we can share them with others.

The next drawing in our Science Notebook shows a measurement in the imperial system and the metric system. During that adventure, we learned that the metric system gave scientists a clear and consistent way to measure and communicate with scientists around the world.

Turn to the next drawing. We learned about mass and matter in our sixth adventure together. Mass measures the amount of material in something or someone, rather than the force of gravity. Mass provides a consistent measurement for scientists. We also learned that just like scientists need a consistent standard to measure things by, we also need a consistent standard for what truth is. What we believe to be truth will be the standard that we live our lives by. God's standard does not change. His Word is truth, and it gives us a firm foundation to build our lives upon.

That's right! In our next adventure, we learned how to describe matter. We also got to pretend to be explorers, like Meriwether Lewis and William Clark. It was a lot of fun! That's all we have time for today, but I can't wait to continue looking through our Science Notebooks with you tomorrow.

1. Look through the first seven drawings in your Science Notebook. Which drawing was your favorite?

______________________________

2. What is something you remember from one of those science adventures?

______________________________

______________________________

______________________________

______________________________

Hey there, friend! We're looking back through all of the drawings in our Science Notebook this week. Isn't it fun to look back and remember all of the different things we've learned about?

Definitely! To get started today, turn to drawing number eight in your Notebook — it's the drawing of an atom! During that adventure, we discovered the building blocks of matter: atoms. We also talked about how the field of chemistry often explores things that we cannot see with our eyes alone. As we study chemistry, we get to see God's infinite wisdom on display. Now, turn to the next page of your Science Notebook — we learned about the elements the following week.

It was so interesting to explore the elements as well as the periodic table of elements that Dmitri Mendeleev developed. Learning about the elements reminded us that as we study God's creation through science, it reveals His eternal power and divine nature to us. The delicate design of a butterfly's wing, incredible night sky, and even the organization of the elements declare to us the glory, power, and majesty of God.

Flip over to the next drawing now. We dove into the periodic table of elements in our tenth adventure together. It was so much fun — which element did you draw, friend? Learning about precious elements, like gold, reminded us that God's ways are firm and righteous. The directions that God gives us in the Bible help us to live our lives the way He designed. They protect us. The Bible is God's Word to us, and it is more precious than gold.

Turn to the next page in your Science Notebook — it's the drawing of a water molecule. We explored molecules in our 11th adventure. A molecule is two or more atoms that have bonded together.

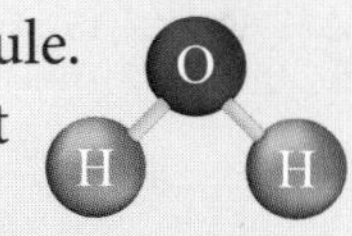


After that, we learned about the carbon cycle. The carbon cycle reminded us that as we study science, we do see things like death and decay that remind us of the consequences of sin. But they also remind us that through Jesus, we don't have to suffer the worst consequence of sin: separation from God. Through Jesus, we can have a relationship with God and live with the hope that one day things will be restored.

In our 13th science adventure, we explored mixtures and solutions; turn to that drawing in your Notebook. Remember, when two or more substances join, or mix together, we call it a mixture. A solution is formed when a substance is dissolved into a different substance.

Name: ______________________________

Our adventure that week also reminded us that the enemy tries to deceive us — just like he did Eve in the Garden — by mixing just a little bit of truth into his lies. In order to know the truth, we must have a relationship with Jesus and study the Word of God, the Bible. The Bible is our standard of truth — and it does not change. The better we know the Bible, the better we become at recognizing truth and lies in the world around us.

We're almost out of time for today, but let's look at one more drawing. Our next drawing was the farm field. We learned about how chemistry is used in the world around us in our 14th adventure together — chemistry is an amazing field of science! See you tomorrow, friend.

apply it

1. Look through the drawings we reviewed today in your Science Notebook. Which drawing was your favorite?

______________________________

______________________________

______________________________

______________________________

2. What is something you enjoyed learning about chemistry?

______________________________

______________________________

______________________________

______________________________

3. What is something that you'd like to learn more about in chemistry?

______________________________

______________________________

______________________________

______________________________

Welcome back! Did you bring your Science Notebook with you, friend? Let's continue looking back through all the things we've learned together. In our 15th adventure, we drew a pair of glasses. Do you remember what we learned that week, Hannah?

I sure do! We began talking about worldview. A worldview is what you believe and the way you see the world around you through your beliefs. In our next adventure, we learned about what shapes our worldview as Christians: the Bible. Flip over to the drawing of your Bible, friend.

After that, we talked about observational and historical science. We drew a picture of a dolphin that week. Observational science is science that we can see, experience, or observe. This type of science is observable, testable, and repeatable.

On the other hand, historical science deals with the past. It cannot be observed, tested, or repeated. As we explore observational and historical science, it's important to remember that the Bible is the foundation we use as Christians to help us interpret evidence and develop theories.

Flip to the next drawing in your Science Notebook — we drew a picture of our favorite living things in our 18th adventure together. During that week, we explored the properties of living things. A living thing is able to breathe, grow, eat or absorb nutrients, move or respond, and reproduce. We also talked about what makes humans different from all other living things. God only made mankind in His image and gave us the breath of life. This is what makes human beings different from all the other living things God created.

Turn to the drawing of a cat now. In this adventure, we learned about how living things are classified through the field of taxonomy. We also talked about how God designed living things to reproduce after their own kind. Even though we may see variations like different colors or sizes within a kind, a dog always stays a dog, a cat always stays a cat, and a duck always stays a duck.

Then we dove right in to the field of marine biology in our 20th adventure together. I had fun drawing a picture of the ocean at the end of that week. After that, we learned about waves and tides together. We drew a picture of the moon to help us remember what we learned that week because the gravitational pull of the moon affects the earth and causes the tide.

Name: ______________________

We began exploring tide pools in our next adventure — turn to your drawing of a sea star. Do you remember how sea stars can regenerate an arm if they lose one to a predator? The word regenerate means to regrow, renew, or to bring about something new. Sea stars reminded us of how God renews and regenerates us. In 2 Corinthians 5:17 it says,

*Therefore, if anyone is in Christ, the new creation has come: The old has gone, the new is here!*

That is one of my favorite verses. I'm excited to continue looking back through our Notebooks with you tomorrow. We explored so much this year and got to see God's wisdom in many different areas of creation!

1. Look through the drawings we reviewed today in your Science Notebook. Which drawing was your favorite?

______________________

2. What is something you enjoyed learning about worldview?

______________________

Hello, friend! Hasn't it been fun to look back through our Science Notebook and remember all of the different things we've learned together? After we learned about tide pools, we created our own model of a tide pool — that was so cool!

It was! Can you find your drawing of a kelp forest, friend? That was one of my favorite lessons. We talked about holdfasts during that adventure. A holdfast is a root-like structure that holds seaweed firmly in place. Without holdfasts, seaweed would be tossed back and forth by the waves. Just as seaweed's holdfast keeps it secure against the wind and waves, the Bible is our holdfast. God's Word holds us secure against the lies of the enemy in the world around us.

Next, we studied the marine food chain and learned that predators have an important purpose in our imperfect world. The food chain we observe today is a sad reminder of the consequences of sin — but it is also a reminder of God's wisdom and mercy even in a broken creation. The food chain helps to keep habitats and ecosystems balanced so that we can enjoy watching and learning about many of God's creations in one place.

Turn to your drawing of a coral reef now, friend. We explored coral reefs in our 26th adventure together. Coral reefs are beautiful ecosystems! The relationships found in the coral reef reminded us that God designed relationships to be an important part of creation — and they are an important part of our lives too. Our relationships with our family and friends are one way we learn more and grow in our faith together.

We learned about the Great Barrier Reef together next. Do you remember learning about the mimic octopus? The mimic octopus can imitate and behave like many different things around it. It can be easy for us to imitate or behave like other people around us too — but this can get us into trouble! Rather than imitate everyone around us, the Bible tells us that we are to be imitators of Christ in Ephesians 5:1,

*Therefore be imitators of God as dear children* (NKJV).

We also created our model of a coral reef at the end of that week. After that, we began learning about whales in our 27th lesson. It sure was fun to draw a picture of the sperm whale in our Notebooks! We continued learning about whales the following week and drew a picture of a blue whale.

Do you remember talking about whale footprints? They reminded us that just as the whale leaves an impact on the water around it, our lives can also have an impact on others. We call the impact our life has on others our influence — and we can have a good or a bad influence on others.

Name:

After that, we talked about conservation and how we can be wise stewards of what God has given us. We each drew a picture of something God has given us to care for. What did you draw, friend?

We're almost done looking through our Science Notebooks now. In our 31st adventure together, we learned about the scientific method in marine biology. I had so much fun drawing the scuba diver! Turn to the drawing after your scuba diver, friend. We learned about sharks next. Whale sharks sure are beautiful.

The following week, we learned about great white sharks, hammerheads, and megalodon. After that, it was time to explore the arctic and create our model biome for the Arctic Ocean.

Whew, we did a lot of learning together and had so much fun exploring God's creation through chemistry and marine biology!

apply it

1. Look through the drawings we reviewed today in your Science Notebook. Which drawing was your favorite?

2. What is something you enjoyed learning about marine biology?

3. What is something you would still like to learn about marine biology?

Well, we've finally made it! We've reached our last science adventure together.

But Hannah, this isn't the end! You've said it before, no matter how much we learn about God's creation, there is always so much more to learn. We still have many, many science adventures ahead of us.

You're right, Ben! I meant to say that we've reached our last adventure together in *this* book. But as we continue to grow, we'll have the opportunity to learn more about chemistry, ecology, geology, biology, physics, and many other fields of science.

And no matter what field of science we're exploring, we can continue to look for God's wisdom, power, majesty, and mercy on display in His creation. In Jeremiah 51:15, we read this about God:

*He made the earth by his power; he founded the world by his wisdom and stretched out the heavens by his understanding.*

Each field of science gives us the opportunity to see the power, wisdom, and understanding of God.

I don't know about you, but the more I learn about science, the more amazed I am by God's power, wisdom, and understanding! Science reminds me of Psalm 111:2–4,

*Great are the works of the LORD; they are pondered by all who delight in them. Glorious and majestic are his deeds, and his righteousness endures forever. He has caused his wonders to be remembered; the LORD is gracious and compassionate.*

Science reveals many of the Lord's great works to us and gives us much to ponder, or think about. I'm so glad we've been able to explore God's creation and learn about the foundation of our worldview together, friend.

As you continue exploring God's creation through science, I pray that you'll continue to look for God's wisdom, power, majesty, and mercy on display. That you will look for evidence of His design in creation as if you were searching for a treasure — and that you will continue sharing what you've learned with others. But most importantly, I pray that you will continue to grow in your relationship with Jesus. That you will build your life on the unchanging truth of God's Word and allow it to shape your worldview.

Now it's time to add the last drawing to our Science Notebook for this year! What do you think we should draw, Ben?

Hmm, let's draw a picture of a treasure chest. Each time we catch a glimpse of God's power, wisdom, and majesty on display in creation through science, it is like finding a treasure.

I like that idea. Here is a picture we can use to give us ideas for our own drawings. Let's get started!

Here is what each of our treasure chests look like. Have fun creating your drawing, friend — and once you're done, don't forget to share it with someone else and tell them all about what you've learned through science.

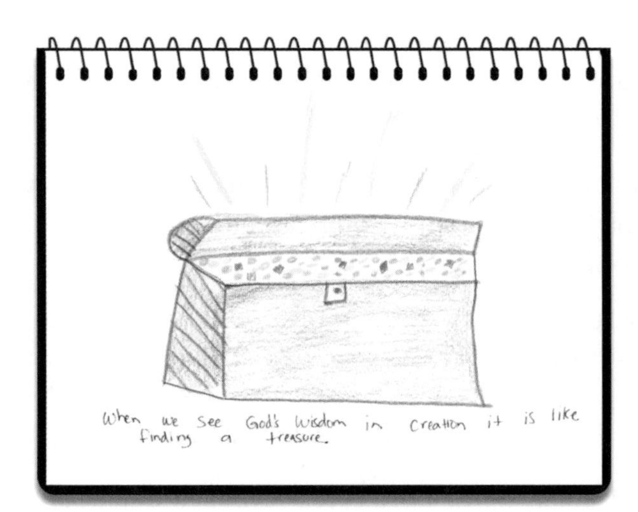

In your Notebook, write:

When we see God's wisdom in creation, it is like finding a treasure!

Then draw a picture of a treasure chest.

Science reveals God's power, wisdom, and understanding to us. Copy Jeremiah 51:15 on the back of your Notebook page as a reminder.

*He made the earth by his power; he founded the world by his wisdom and stretched out the heavens by his understanding* (Jeremiah 51:15).

# Certified Science Explorer

This award is presented to ______________________________
(Student's Name)

for successfully completing *Adventures in the Scientific Method*.

This certifies that ______________________________ is
(Student's Name)

ready to look for God's wisdom, power, majesty, and mercy on display in His creation as they continue to explore science in years to come.

Teacher: ______________________________

Date: ____________________

# PERIODIC TABLE OF ELEMENTS

| | | | | | | | | | | | | | | | | | |
|---|---|---|---|---|---|---|---|---|---|---|---|---|---|---|---|---|---|
| 1 H Hydrogen | | | | | | | | | | | | | | | | | 2 He Helium |
| 3 Li Lithium | 4 Be Beryllium | | | | | | | | | | | 5 B Boron | 6 C Carbon | 7 N Nitrogen | 8 O Oxygen | 9 F Fluorine | 10 Ne Neon |
| 11 Na Sodium | 12 Mg Magnesium | | | | | | | | | | | 13 Al Aluminium | 14 Si Silicon | 15 P Phosphorus | 16 S Sulfur | 17 Cl Chlorine | 18 Ar Argon |
| 19 K Potassium | 20 Ca Calcium | 21 Sc Scandium | 22 Ti Titanium | 23 V Vanadium | 24 Cr Chromium | 25 Mn Manganese | 26 Fe Iron | 27 Co Cobalt | 28 Ni Nickel | 29 Cu Copper | 30 Zn Zinc | 31 Ga Gallium | 32 Ge Germanium | 33 As Arsenic | 34 Se Selenium | 35 Br Bromine | 36 Kr Krypton |
| 37 Rb Rubidium | 38 Sr Strontium | 39 Y Yttrium | 40 Zr Zirconium | 41 Nb Niobium | 42 Mo Molybdenum | 43 Tc Technetium | 44 Ru Ruthenium | 45 Rh Rhodium | 46 Pd Palladium | 47 Ag Silver | 48 Cd Cadmium | 49 In Indium | 50 Sn Tin | 51 Sb Antimony | 52 Te Tellurium | 53 I Iodine | 54 Xe Xenon |
| 55 Cs Caesium | 56 Ba Barium | 57 La* Lanthanum | 72 Hf Hafnium | 73 Ta Tantalum | 74 W Tungsten | 75 Re Rhenium | 76 Os Osmium | 77 Ir Iridium | 78 Pt Platinum | 79 Au Gold | 80 Hg Mercury | 81 Tl Thallium | 82 Pb Lead | 83 Bi Bismuth | 84 Po Polonium | 85 At Astatine | 86 Rn Radon |
| 87 Fr Francium | 88 Ra Radium | 89 Ac** Actinium | 104 Rf Rutherfordium | 105 Db Dubnium | 106 Sg Seaborgium | 107 Bh Bohrium | 108 Hs Hassium | 109 Mt Meitnerium | 110 Ds Darmstadtium | 111 Rg Roentgenium | 112 Cn Copernicium | 113 Nh Nihonium | 114 Fl Flerovium | 115 Mc Moscovium | 116 Lv Livermorium | 117 Ts Tennessine | 118 Og Oganesson |

| | | | | | | | | | | | | | | |
|---|---|---|---|---|---|---|---|---|---|---|---|---|---|---|
| * | 58 Ce Cerium | 59 Pr Praseodymium | 60 Nd Neodymium | 61 Pm Promethium | 62 Sm Samarium | 63 Eu Europium | 64 Gd Gadolinium | 65 Tb Terbium | 66 Dy Dysprosium | 67 Ho Holmium | 68 Er Erbium | 69 Tm Thulium | 70 Yb Ytterbium | 71 Lu Lutetium |
| ** | 90 Th Thorium | 91 Pa Protactinium | 92 U Uranium | 93 Np Neptunium | 94 Pu Plutonium | 95 Am Americium | 96 Cm Curium | 97 Bk Berkelium | 98 Cf Californium | 99 Es Einsteinium | 100 Fm Fermium | 101 Md Mendelevium | 102 No Nobelium | 103 Lr Lawrencium |

| METAL | | | | | | METALLOID | NONMETAL | | UNKNOWN |
|---|---|---|---|---|---|---|---|---|---|
| Alkali metal | Alkaline earth metal | Lanthanide | Actinide | Transition metal | Post-transition metal | Metalloid | Reactive nonmetal | Noble gas | unknown |

# Basic Phonics Review

Phonetic pronunciations are included throughout this book. Your student may review the vowel symbols and pronunciation using this chart.

| Vowel | As In |
|---|---|
| ă | dad |
| ā | cape |
| å | far |

| Vowel | As In |
|---|---|
| ĕ | men |
| ē | be |
| ĭ | sit |
| ī | like |

| Vowel | As In |
|---|---|
| ŏ | not |
| ō | bone |
| ŭ | sun |
| ū | use |

# Glossary

## A

**Abbreviation** (said this way: ŭh-brē-vē-ā-shŭhn): a shorter way to write a word.

**Adapt** (said this way: ŭh-dăpt): to adjust or change for certain conditions or a particular environment.

**Analyze** (said this way: ăn-l-līze): to closely examine.

**Anatomy** (said this way: ŭh-năt-ŭh-mē): parts of a body.

**Apex** (said this way: āpĕx): means the top.

**Archaeology** (said this way: årk-ē-ŏl-ŭh-jē): a field of science that studies historical events.

**Atomic number** (said this way: : ŭh-tŏm-ĭk nŭhm-ber): tells us how many protons an element has in its nucleus.

## B

**Bias** (said this way: bī-ŭhs): a belief, opinion, or worldview that shapes how we see the world.

**Buoyancy** (said this way: boy-ŭhn-sē): the ability to float or be suspended by water.

## C

**Camouflage** (said this way: kă-mŭh-flåzh): a way to stay hidden in an environment.

**Carnivore** (said this way: kår-nĕ-vōr): a consumer that only eats other consumers like fish or animals.

**Characteristic** (said this way: kār-ĭk-ter-ĭst-ĭk): a trait or feature of something.

**Classify** (said this way: klăs-ŭh-fī): to organize or arrange things.

**Compound** (said this way: kŏm-pound): two or more different atoms bonded together to form a molecule.

**Conclusion** (said this way: kŭhn-kloo-zhŭn): a result or decision about something.

**Conservation** (said this way: kŏn-ser-vāy-shŭn): the work we do to care for, protect, and preserve God's creation.

**Consumer** (said this way: kŭn-soo-mer): isn't able to make its own food like a plant can; it must consume food.

**Creationism** (said this way: krē-ā-shŭh-nĭz-ŭhm): the theory that God created the heavens, the earth, and all living things, just as it says in Genesis.

**Cubit** (said this way: kyoo-bĭt): the length from an adult's elbow to the tip of their middle finger.

**Cycle** (said this way: sī-kŭhl): a process or series of events that happens over and over again.

## D

**Decay** (said this way: dē-kā): the process of something breaking down.

**Decomposition** (said this way: dē-cŏm-pō-zĭ-shŭn): the process in which dead plants and animals break back down into the soil.

**Definition** (said this way: dĕf-ŭh-nĭsh-ŭhn): what a word means.

**Density** (said this way: dĕn-sĭ-tē): the measurement of how much matter is in an object and how much space that matter takes up.

**Diatomic** (said this way: dī-ŭh-tŏm-ĭk): two atoms bonded together form a diatomic molecule.

**Dissolve** (said this way: dĭh-zŏlv): a word that means to melt or mix something into something else.

## E

**Echolocation** (said this way: ĕk-ōh-lōh-kāy-shŭhn): a feature that allows bats, whales, and dolphins to use sound waves in order to "see" what is around them.

**Ecosystem** (said this way: ē-cō-sĭs-tŭm): a community of living and nonliving things that are together in one place.

**Ectothermic** (said this way: ĕk-tōh-thur-mĭc): cold-blooded.

**Element** (said this way: ĕl-ŭh-mĕnt): a pure substance that cannot be broken down into any other substance.

**Empirical** (said this way: ĕm-pir-ĭ-kŭhl): means that something can be supported or verified through an experience or an experiment.

**Endothermic** (said this way: ĕn-dōh-thur-mĭc): warm-blooded.

**Evidence** (said this way: ĕv-ĭ-dŭhns): the information we can see or experience around us.

**Exoskeleton** (said this way: ĕk-sō-skĕl-ĭ-tŏn): a hard shell that functions as a skeleton on the outside of a creature's body.

**Extinct** (said this way: ĭk-stingkt): a species of living thing is no longer found alive anywhere on the earth.

## F

**Fertilize** (said this way: fur-tĭl-īz): to apply something to the soil to add nutrients or make the soil richer.

**Friction** (said this way: frĭk-shŭhn): a type of force that pushes against movement.

## G

**Geocentrism** (said this way: jē-ō-sĕn-trĭsm): the theory that the earth is at the center of the universe.

**Glacier** (said this way: glāy-shur): a very large area of ice.

## H

**Habitat** (said this way: hăb-ĭ-tăt): the natural environment a plant or animal lives in.

**Heliocentrism** (said this way: hē-lē-ō-sĕn-trĭsm): the theory that the sun is at the center of the universe.

**Herbivore** (said this way: er-bĭh-vōr): a consumer that only eats primary producers like plants and algae.

**Heterogeneous mixture** (said this way: hĕt-ŭh-rå-jŭh-nŭhs mĭx-cher): a mixture in which the substances in the mixture aren't evenly mixed together or distributed.

**Homogenous mixture** (said this way: hŭh-mŏj-ŭh-nŭhs mĭx-cher): a mixture in which the substances are evenly distributed throughout the whole mixture.

**Horizontal** (said this way: hōr-ŭh-zŏn-tl): means that something is in a side to side position.

**Hypothesis** (said this way: hī-pŏth-ŭh-sĭs): explains what may be happening or gives a possible answer for the questions we've asked. A hypothesis is something that we can test through an experiment.

## I

**Imperial** (said this way: ĭm-pēr-ē-ŭhl): the imperial system can also be called the English system, and it is the system most commonly used to measure things in the United States.

**Interpret** (said this way: ĭn-tur-prĭt): to explain what something means or to develop an understanding.

**Intertidal zone** (said this way: ĭn-ter-tī-dåhl zōne): the area that is visible during low tide but under water during high tide.

**Invertebrate** (said this way: in-vur-tŭh-brŭht): a living thing without a backbone.

## K

**Keystone species** (said this way: kēē-stōne spē-shēz): a type of animal that the ecosystem depends upon.

**Kind** (said this way: kīnd): a group of living things that have similar traits and can reproduce.

## L

**Lagoon** (said this way: lŭh-goon): an area of still, quiet ocean water that is separated from the ocean by a sandbank or island.

**Logical** (said this way: lŏj-ĭ-kŭhl): something that would be expected or makes sense.

## M

**Mass** (said this way: măs): measures the amount of material in something or someone, rather than the force of gravity.

**Matter** (said this way: măt-er): anything that takes up space and has mass.

**Metric** (said this way: mĕ-trĭk): was designed to be a clear and consistent way to measure things. It uses multiples of 10.

**Migrate** (said this way: mīgrāt): means to travel from one place to another, often over a very large distance.

**Mixture** (said this way: mĭx-cher): two or more substances joined, or mixed together.

**Model** (said this way: mŏd-l): an example of what something looks like.

**Molecule** (said this way: mŏl-ŭh-kyool): two or more atoms combined, or bonded together.

**Molting** (said this way: mōlt-ing): the process in which a living thing sheds old skin, feathers, or hair.

**Mutualism** (said this way: myoo-choo-ŭh-lĭz-ŭhm): a symbiotic relationship in which two organisms receive a benefit from each other.

## N

**Nucleus** (said this way: noo-klē-ŭhs): a group of protons and neutrons together.

**Nutrients** (said this way: new-trēē-ĕnt): a substance that plants, animals, and people need to grow and live. Nutrients provide the energy that living things need.

## O

**Observe** (said this way: ŭhb-zerv): to see or notice something.

**Omnipotent** (said this way: ŏmnĭp-ŭh-tĕnt): a word that means all mighty, with no limit.

**Omniscient** (said this way: ŏm-nĭsh-ŭhnt): a word meaning that God knows everything.

**Omnivore** (said this way: åm-nĭh-vōr): a consumer that eats both plants and other creatures.

## P

**Predator** (said this way: prĕ-dŭh-ter): a consumer that hunts and eats other animals.

**Prefix** (said this way: prē-fĭx): a word placed in front of another word, like "centi" in centimeter.

**Prey** (said this way: prāy): an animal that is hunted and eaten by another animal.

**Primary producers**: produce their own food for energy. In the ocean, algae, seaweed, and seagrass would be examples of primary producers.

**Property** (said this way: prŏp-er-tē): a trait or feature something has.

**Pursue** (said this way: per-soo): to chase after, seek, or search for something.

## R

**Repel** (said this way: rĭ-pĕl:) to push something away.

**Respiration** (said this way: rĕs-pŭh-rāy-shŭn): the process of breathing.

## S

**Science**: the pursuit of knowledge and understanding about God's creation through an organized process. Science helps us to ask questions, test our ideas, and share what we've learned with others. Through science, we also learn more about God and our relationship with Him.

**Solute** (said this way: sŏl-yoot): the substance that is dissolved in another substance.

**Solution** (said this way: sŭh-loo-shŭhn): formed when a substance is dissolved into a different substance.

**Solvent** (said this way: sŏl-vŭhnt): the substance that dissolves another substance.

**Structure** (said this way: strŭk-cher): provides a way to organize or support something.

**Substance** (said this way: sŭb-stŭhns): a certain kind of matter.

**Symbiotic relationship** (said this way: sĭm-bē-ŏt-ĭk): organisms living together in a close relationship. This is also called symbiosis (said this way: sĭm-bē-ōh-sĭs).

## T

**Taxonomy** (said this way: tăk-sŏn-ŭh-mē): the field of science that deals with classifying living things — and it's part of biology!

**Theory** (said this way: thē-ŭh-rē): a logical way to explain what we see or to answer a question based on evidence and facts. Once a theory has been developed, a scientist can continue working to support the theory with additional evidence.

## V

**Vertebrates** (said this way: vur-tŭh-brŭhts): living things that have a backbone.

**Vertical** (said this way: ver-tĭ-kŭhl): means that something is in a position of being up and down.

## W

**Worldview** (said this way: wurld-vyoo): what you believe and the way you see the world around you through your beliefs.

# Answer Keys

## Page 20

1. Answer will vary but should include something the student has enjoyed learning about science in the past.
2. Answer will vary but should include something the student wants to learn about in science.
3. A definition is what a word means.

## Page 22

1. Science is the pursuit of knowledge and understanding about God's creation through an organized process. Science helps us to ask questions, test our ideas, and share what we've learned with others. Through science, we also learn more about God and our relationship with Him.
2. Definitions may vary based on the dictionary used.

## Page 24

1. knowledge, understanding, creation, ask, test, share, God
2. Answer will vary, but student should list some fields of science they have heard about.
3. Answer will vary, but student should include the name of a scientist they have learned about in the past and their field of study.

## Page 30

1. Answer will vary.
2. Answer will vary.

## Page 32

All answers will vary.

## Page 34

1. Heliocentrism — The theory that the sun is at the center of the solar system and that the earth and planets orbit the sun.

   Geocentrism — The theory that the earth is at the center of the solar system and that the sun and planets orbit the earth.

2. Answers will vary, but student should explain why their five senses are or are not always trustworthy.

## Page 38

Francis Bacon is recognized as the man who developed what we call the scientific method.

## Page 40

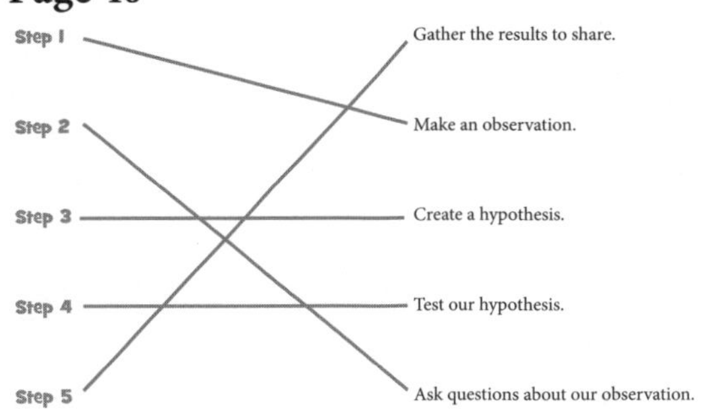


## Page 42

Experiment answers will vary.

## Page 48

Obervation answers will vary.

## Page 50

1. Answers will vary but may include: Lab reports help us to record and share information, questions, our hypothesis, and experiments.
2. Answers will vary but should include how a lab report can help us as we continue to explore science.

## Page 60

Student should explain why they think it is important for measurements to stay consistent. All other answers will vary.

## Page 62

Measurements will vary.

## Page 64

All answers will vary.

## Page 70

Student should explain why they think it is important for measurements to stay consistent. All other answers will vary.

## Page 72

All answers will vary.

## Page 82

All answers will vary.

## Page 84

Observation answers will vary.

## Page 92

All answers will vary.

## Page 94

Experiment answers will vary.

## Page 100

1. Answer will vary.
2. 47
3. Answer will vary.

## Page 104

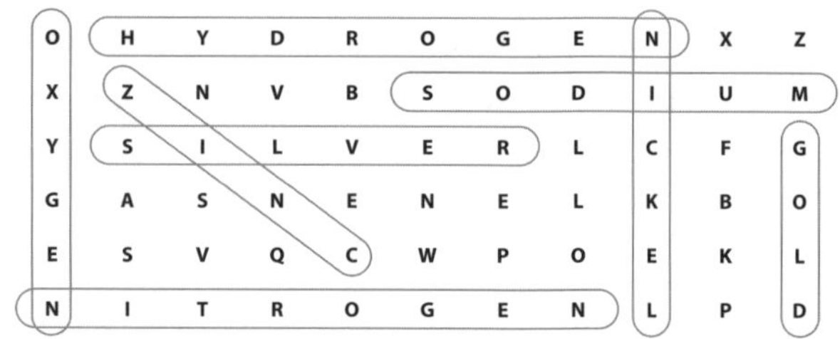


## Page 110

Student should write names of elements they've learned or heard about before. Other answers will vary.

## Page 112

1.

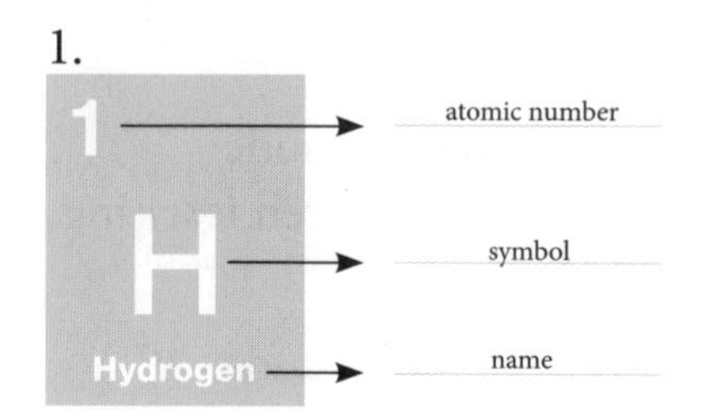


2. Answer may vary but should include that these gases don't react to other elements.

## Page 114

1. Copper, 29
2. Aluminum, 13
3. Oxygen, 8
4. Iron, 26
5. Nickel, 28

## Page 118

1. diatomic
2. molecule

## Page 124

2. Answers may vary but should include that the number tells us the number of atoms.
3. b
4. $H_2O_2$

## Page 128

1. Transpiration
2. Cycle
3. Evaporates
4. Answers may vary but should include water would be used up.

## Page 132

All answers will vary.

## Page 138

1. A mixture is two or more substances mixed together.

2. Answers may vary, may include chocolate milk, tea (sugar, milk, water), baking, etc.
3. Does not
4. Can
5. Can

**Page 140**

1. Cup number 1 was evenly mixed.
2. Yes
3. Cup number 2 was unevenly mixed; the oil stayed on the surface.
4. No
5. Homogenous
6. Heterogenous

**Page 148**

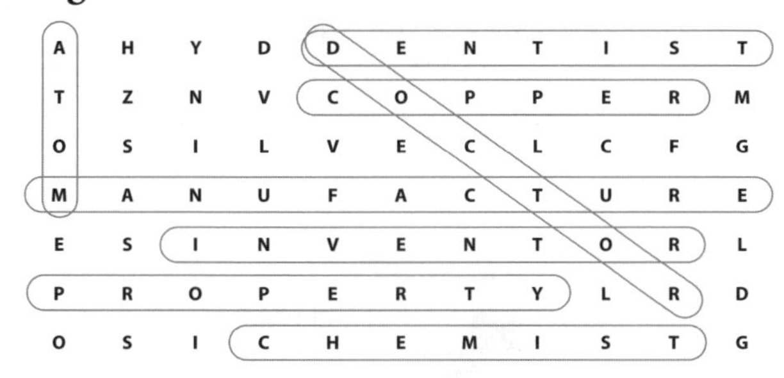


**Page 150**

1. Carbon, hydrogen, nitrogen, oxygen
2. Carbon, hydrogen, nitrogen, oxygen
3. Carbon, hydrogen, nitrogen, oxygen
4. Answers may vary, but may include that there are different amounts of atoms in each formula.

**Page 158**

1. knowledge, understanding, organized process, ask questions, ideas, others, God, relationship
2. Answers will vary, but student should include things they think could shape the way we study science.

**Page 160**

All answers will vary.

**Page 162**

1. A worldview is what you believe and the way you see the world around you through your beliefs.
2. The Bible
3. God confused the languages at the Tower of Babel.

**Page 166**

All answers will vary.

**Page 170**

1. A bias is a belief, opinion, or worldview that shapes how we see the world.
2. Answers will vary, but students should include what they believe about the Bible.

**Page 174**

1. Observable, testable, repeatable
2. Supported, verified, experience, experiment
3. Answers will vary, but student should include something they would like to study or observe.
4. Answers will vary, but student should include how they would study or observe what they chose in question 3 like a scientist.

**Page 176**

1. Land-dwelling creatures, mankind
2. Seed-bearing plants
3. Answers may vary but should include that observational science can be observed, tested, and repeated but historical science cannot.

**Page 178**

1. Can this information be observed, tested, and repeated?

2. Observational science
3. Historical science
4. Observational science

**Page 182**

Note: terms may vary by Bible translation.

1. Heavens, earth, light, water
2. Sky/atmosphere, divided water
3. Dry land, plants
4. Sun, moon, stars
5. Sea creatures, flying creatures
6. Land creatures, mankind
7. God rested.
8. Answers may vary but should include that God created living things.

**Page 184**

1. Answer will vary.
2. Answer will vary, but student should list other ways they think we might be able to identify living from nonliving things.

**Page 186**

1–5, wording and order may vary: respiration/breathing, grow, reproduce, move/respond, eat/absorb nutrients

**Page 196**

1. Animal, vertebrate
2. Answers will vary, but student should describe what the backbone feels like.
3. Humans are made in the image of God.

**Page 198**

| | | | | | | | | | | | |
|---|---|---|---|---|---|---|---|---|---|---|---|
| A | V | E | R | T | E | B | R | A | T | E | M |
| C | A | C | A | R | N | I | V | O | R | E | A |
| A | N | I | M | A | L | C | O | E | J | K | M |
| M | A | N | U | A | A | C | T | U | R | E | M |
| L | S | I | T | V | E | N | T | O | R | Z | A |
| P | R | I | D | O | M | E | S | T | I | C | L |
| I | N | V | E | R | T | E | B | R | A | T | E |

**Page 202**

1. A marine biologist studies the ocean and the life it contains.
2. Answers will vary, but student should describe what they think it would be like to be a marine biologist.
3. Answers will vary, but student should include something they would like to learn about the ocean or marine life.

**Page 204**

Observation answers will vary.

**Page 212**

Observation answers will vary.

**Page 216**

1. Answers may vary but should include the moon or the moon's gravitational pull on the earth.
2. 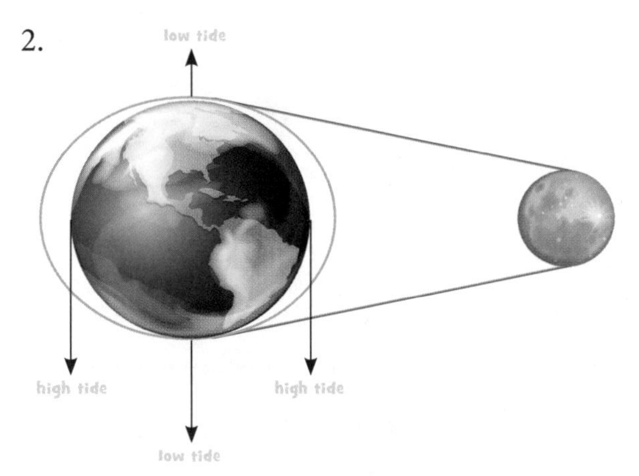


**Page 220**

2. The intertidal zone
3. A community of living and nonliving things that are together in one place.

**Page 222**

1–2. Answers will vary.

3. To be able to adjust or change for certain conditions or a particular place.
4. Answers may vary but should include that conditions are always changing.

**Page 236**

1. Plant, Animal, Protista
2. Protista
3. Answers may vary, should include that algae produces oxygen, even more oxygen than plants.

4–5. Answers will vary.

**Page 238**

Answers will vary.

**Page 244**

1. Answers will vary, but student should describe what they think it would have been like to live in God's original creation where animals and people ate only plants.
2. Answers may vary depending on the Bible translation but should include that the fear and dread of people would fall on the animals.
3. Answers may vary, but student should include what they think it would have been like to see a lion, tiger, bear, or dinosaur before it was afraid of people.

**Page 246**

1. Primary producers
2. Omnivore
3. Herbivore
4. Carnivore

**Page 254**

1. Fringing reef
2. Atoll
3. Barrier reef

**Page 258**

1. Answers may vary but should include: A relationship in which two organisms receive a benefit from each other.
2. Answers will vary, but student should include other relationships in creation where two living things receive a benefit from each other (e.g., bees and flowers).
3. Answers will vary, but student should include why they think God may have designed mutualistic relationships in creation.

**Page 266**

1. Answers will vary, but student should include what similarities they see between the species of sea turtles.
2. Answers will vary, but student should include what differences they notice between the species of sea turtles.
3. Answers will vary, but student should include how they think the green sea turtle is able to migrate such long distances.
4. Answers will vary, but student should include how they might test their theory of turtle migration if they were a marine biologist.

**Page 270**

1. Answers may vary.
2. Hair on their bodies, feed their babies milk, need oxygen to breathe, and are vertebrates.
3. Humans are made in the image of God.

**Page 272**

1. The parts of a body.
2. Answers may vary but should include that a whale's rib cage is flexible, which allows it to dive deep in the ocean.

**Page 274**

1. Answer will vary, but student should have written down how long they can hold their breath.
2. Answer will vary, but student should include how they think a sperm whale is able to hold its breath for 45 minutes.

3. Answer will vary, but student should include what they thought was the most interesting thing about sperm whales.

4. Answer will vary, but student should include what they would want to learn about sperm whales if they were a marine biologist.

**Page 286**

All answers will vary.

**Page 292**

1. Conservation means to protect or keep from wasting something.

2. Answer will vary, but student should list something God has placed in their care.

3. Answer will vary, but student should include how they care for what they wrote in question 2 as a wise steward.

**Page 294**

1. Translation may vary: "Everything that lives and moves about will be food for you. Just as I gave you the green plants, I now give you everything."

2. The species is no longer allowed to be hunted, or it can only be hunted under certain conditions.

3. Answers may vary. Examples include elephants, tigers, gorillas, pandas, jaguars, leopards, etc.

**Page 296**

1. Answers will vary, but student should include their favorite thing to learn about in God's creation.

2. Answers will vary, but student should include what they would like to learn more about in God's creation.

3. Answers may vary but could include taking care of your room, belongings, or pets; learning about God's creation; conserving resources; etc.

**Page 300**

Observation answers will vary.

**Page 302**

1–2. Answers will vary.

3. Answer will vary, but student should include a question they would have about bottlenose dolphins.

4. Answer will vary, but student should include how they think they could answer their question from number 3.

5. Answer will vary, but student should include what they think it would be like to study bottlenose dolphins for over 50 years.

**Page 304**

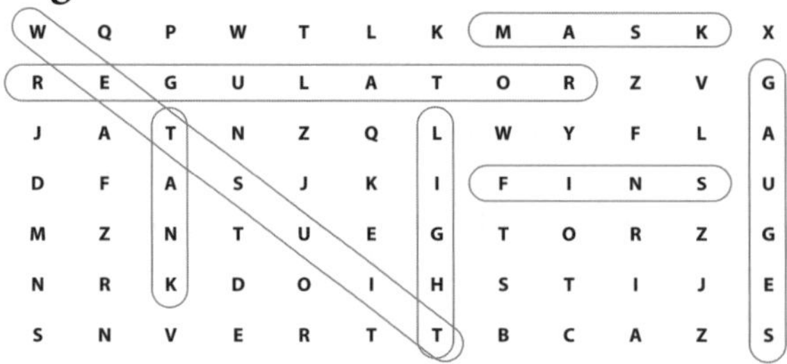


**Page 310**

1. Cold-blooded

2. Warm-blooded

3. Vertebrate, cold-blooded, fins, gills, lives in water

**Page 314**

1. Answer will vary, but student should include how they would describe a whale shark.

2–3. Answers will vary, but student should include why they think there are or are not other living creatures in the ocean that we haven't discovered yet.

**Page 322**

Descriptions will vary.

**Page 330**

1. Answer will vary, but student should include something they would like to learn about the Arctic Ocean.
2. Answer will vary but may include because of the thick ice and frigid temperatures.

**Page 340**

1. Answers may vary but may include: have long claws to dig through the ice, are excellent swimmers, have a thick layer of fat or blubber.
2. By blowing bubbles underneath the hole.

**Page 342**

1. Answers will vary but may include the following information about ringed seals: excellent swimmers, hold breath for 45 minutes, have claws to dig through ice, maintain several breathing holes, blow bubbles to check for polar bears, eat arctic cod, have a thick layer of blubber.
2. They keep the fish from freezing.

**Page 344**

1. Bowhead whale
2. Polar bear
3. Polar cod
4. Greenland shark
5. Beluga whale
6. Ringed seal

**Page 348**

1. Answer will vary.
2. Answer will vary, but student should include something they remember from one of the first seven science adventures.

**Page 350**

1. Answer will vary.
2. Answer will vary, but student should include something they enjoyed learning about chemistry.
3. Answer will vary, but student should include something they'd like to learn more about in chemistry.

**Page 352**

1. Answer will vary.
2. Answer will vary, but student should include something they enjoyed learning about worldview.

**Page 354**

1. Answer will vary.
2. Answer will vary, but student should include something they enjoyed learning about marine biology.
3. Answer will vary, but student should include something they would still like to learn about marine biology.